Nuclear Physics

The Core of Matter, The Fuel of Stars

Committee on Nuclear Physics
Board on Physics and Astronomy
Commission on Physical Sciences, Mathematics, and Applications
National Research Council

NATIONAL ACADEMY PRESS
Washington, D.C. 1999

NOTICE: The project that is the subject of this report was approved by the Governing Board of the National Research Council, whose members are drawn from the councils of the National Academy of Sciences, the National Academy of Engineering, and the Institute of Medicine. The members of the committee responsible for the report were chosen for their special competences and with regard for appropriate balance.

The National Academy of Sciences is a private, nonprofit, self-perpetuating society of distinguished scholars engaged in scientific and engineering research, dedicated to the furtherance of science and technology and to their use for the general welfare. Upon the authority of the charter granted to it by Congress in 1863, the Academy has a mandate that requires it to advise the federal government on scientific and technical matters. Dr. Bruce Alberts is president of the National Academy of Sciences.

The National Academy of Engineering was established in 1964, under the charter of the National Academy of Sciences, as a parallel organization of outstanding engineers. It is autonomous in its administration and in the selection of its members, sharing with the National Academy of Sciences the responsibility for advising the federal government. The National Academy of Engineering also sponsors engineering programs aimed at meeting national needs, encourages education and research, and recognizes the superior achievements of engineers. Dr. William A. Wulf is president of the National Academy of Engineering.

The Institute of Medicine was established in 1970 by the National Academy of Sciences to secure the services of eminent members of appropriate professions in the examination of policy matters pertaining to the health of the public. The Institute acts under the responsibility given to the National Academy of Sciences by its congressional charter to be an advisor to the federal government and, upon its own initiative, to identify issues of medical care, research, and education. Dr. Kenneth I. Shine is president of the Institute of Medicine.

The National Research Council was established by the National Academy of Sciences in 1916 to associate the broad community of science and technology with the Academy's purposes of furthering knowledge and advising the federal government. Functioning in accordance with general policies determined by the Academy, the Council has become the principal operating agency of both the National Academy of Sciences and the National Academy of Engineering in providing services to the government, the public, and the scientific and engineering communities. The Council is administered jointly by both Academies and the Institute of Medicine. Dr. Bruce Alberts and Dr. William A. Wulf are chairman and vice chairman, respectively, of the National Research Council.

This project was supported by the Department of Energy under Contract No. DE-FG02-96ER40957 and the National Science Foundation under Grant No. PHY-9515524. Any opinions, findings, conclusions, or recommendations expressed in this publication are those of the authors and do not necessarily reflect the views of the agencies that provided support for this project.

Library of Congress Card Catalog Number 98-89539

International Standard Book Number 0-309-06276-4

Additional copies of this report are available from:

National Academy Press (http://www.nap.edu)
2101 Constitution Ave., NW, Box 285
Washington, D.C. 20055
800-624-6242
202-334-3313 (in the Washington metropolitan area)

Board on Physics and Astronomy
National Research Council, HA 562
2101 Constitution Avenue, N.W.
Washington, DC 20418

Printed in the United States of America

COMMITTEE ON NUCLEAR PHYSICS

JOHN P. SCHIFFER, Argonne National Laboratory and University of Chicago, *Chair*
SAM M. AUSTIN, Michigan State University
GORDON A. BAYM, University of Illinois at Urbana-Champaign
THOMAS W. DONNELLY, Massachusetts Institute of Technology
BRADLEY FILIPPONE, California Institute of Technology
STUART FREEDMAN, University of California at Berkeley
WICK C. HAXTON, University of Washington
WALTER F. HENNING, Argonne National Laboratory
NATHAN ISGUR, Thomas Jefferson National Accelerator Facility
BARBARA JACAK, State University of New York at Stony Brook
WITOLD NAZAREWICZ, University of Tennessee at Knoxville
VIJAY R. PANDHARIPANDE, University of Illinois at Urbana-Champaign
PETER PAUL,* State University of New York at Stony Brook
STEVEN E. VIGDOR, Indiana University

DONALD C. SHAPERO, Director
ROBERT L. RIEMER, Senior Program Officer

---

*Currently at Brookhaven National Laboratory.

COMMISSION ON PHYSICAL SCIENCES, MATHEMATICS, AND APPLICATIONS

Preface

The Committee on Nuclear Physics was established by the Board on Physics and Astronomy as part of its decadal survey series, *Physics in a New Era*. The committee met four times over the course of a year. It heard from program managers at the U.S. Department of Energy (DOE) and the National Science Foundation (NSF) and solicited input from the nuclear physics community through the American Physical Society's Division of Nuclear Physics. A set of peer readers who were asked by the committee to read the draft report (J. Friar, G. Garvey, K. Gelbke, E. Henley, B. Holstein, and R. Holt) provided valuable perspectives, and their comments had an influence on the report. The comments of the reviewers of the report (see page ix) also provided useful input. The committee would like to thank both groups for their time and help. In addition, the list of individuals who helped with material for the report in a variety of ways is too long to enumerate, and the committee wishes to express its gratitude for this assistance.

As part of the physics survey, the overall objective of the study was to help the general public, the government agencies concerned with the support of science, Congress, and the physics community to envision the future of this field within the nation's overall physics effort.

The report of the committee is in the context of previous planning for the field and follows the reports of the 1972 Physics Survey's Nuclear Physics Panel, chaired by J. Weneser, and the 1986 Physics Survey's Nuclear Physics Panel, chaired by J. Cerny, as well as the planning of the Nuclear Science Advisory Committee (NSAC), a joint advisory committee of the NSF and DOE. In particular, the committee drew on the 1996 report *Nuclear Science: A Long Range Plan*, prepared by NSAC (available from the Division of Nuclear Physics, Office of Science, DOE, or the Nuclear Science Section, Physics Division, NSF). The NSAC Long Range Plans represent a wide community involvement in shaping the field.

The committee would like to acknowledge the assistance of Donald C. Shapero, director, Board on Physics and Astronomy, and Robert L. Riemer, senior program officer. The committee would especially like to acknowledge the late David Schramm, who started the present review and discussed the committee's task at its first meeting. His unique enthusiasm for physics and profound interest and many contributions to the nuclear physics of astrophysical phenomena are missed in nuclear physics as indeed in all of physics.

The committee would like to acknowledge the support provided by grants from NSF's Physics Division and DOE's Office of Science.

John P. Schiffer, *Chair*
Committee on Nuclear Physics

Acknowledgment of Reviewers

This report has been reviewed by individuals chosen for their d
spectives and technical expertise, in accordance with procedures appr(
National Research Council's (NRC's) Report Review Committee. T
of this independent review is to provide candid and critical commen
assist the authors and the NRC in making the published report a:
possible and to ensure that the report meets institutional standards for (
evidence, and responsiveness to the study charge. The contents of
comments and draft manuscript remain confidential to protect the inte;
deliberative process. We wish to thank the following individuals for th
pation in the review of this report:

Gary Adams, Rensselaer Polytechnic University,
Felix Boehm, California Institute of Technology,
Stanley J. Brodsky, Stanford Linear Accelerator Center,
Richard Casten, Yale University,
Ernest M. Henley, University of Washington,
Jerry Garrett, Oak Ridge National Laboratory,
J. Ross Macdonald, University of North Carolina,
Peter Parker, Yale University,
R.G. Hamish Robertson, University of Washington, and
James P. Vary, Iowa State University.

Although the individuals listed above have provided many constru
ments and suggestions, responsibility for the final content of this r
solely with the authoring committee and the NRC.

Contents

Nuclear Physics

Summary and Recommendations

Nuclear physics addresses the nature of matter making up 99.9 percent of the mass of our everyday world. It explores the nuclear reactions that fuel the stars, including our Sun, which provides the energy for all life on Earth. The field of nuclear physics encompasses some 3,000 experimental and theoretical researchers who work at universities and national laboratories across the United States, as well as the experimental facilities and infrastructure that allow these researchers to address the outstanding scientific questions facing us. This report provides an overview of the frontiers of nuclear physics as we enter the next millennium, with special attention to the state of the science in the United States.

The current frontiers of nuclear physics involve fundamental and rapidly evolving issues. One is understanding the structure and behavior of strongly interacting matter in terms of its basic constituents, quarks and gluons, over a wide range of conditions—from normal nuclear matter to the dense cores of neutron stars, and to the Big Bang that was the birth of the universe. Another is to describe quantitatively the properties of nuclei, which are at the centers of all atoms in our world, in terms of models derived from the properties of the strong interaction. These properties include the nuclear processes that fuel the stars and produce the chemical elements. A third active frontier addresses fundamental symmetries of nature that manifest themselves in the nuclear processes in the cosmos, such as the behavior of neutrinos from the Sun and cosmic rays, and in low-energy laboratory tests of these symmetries.

With recent developments on the rapidly changing frontiers of nuclear physics the Committee on Nuclear Physics is greatly optimistic about the next ten years. Important steps have been taken in a program to understand the structure of matter in terms of quarks and gluons. The United States has made two major and farsighted investments in this program. The Continuous Electron Beam Accelerator Facility (CEBAF) has recently come into operation and is now delivering beams of unprecedented quality. It will serve as the field's primary "microscope" for probing the building blocks of matter such as the nucleons (protons, neutrons) and the nuclei of atoms, at the small length scales where new physics phenomena involving quarks and gluons should first appear. It will provide new insights into the structure of both isolated nucleons and nucleons imbedded in the nuclear medium. The Relativistic Heavy Ion Collider (RHIC), whose construction is now nearing completion, will produce the world's most energetic collisions of heavy nuclei. This will allow nuclear physicists to probe the properties of matter at energies and densities similar to those characterizing the cores of neutron stars and the Big Bang. RHIC experiments should teach us about the expected transition to a new phase of nuclear matter in which the quarks and gluons are no longer confined within nucleons and mesons.

The theory supporting these new efforts has produced new bridges between quantum chromodynamics (QCD)—the theory of quarks and gluons—and the field's more traditional models of nuclear structure, which involve nucleons and mesons. Nuclear theorists have begun to construct "effective theories" that are equivalent to QCD at low energies, yet share many of the properties of traditional models that view nuclei as quantum fluids of protons and neutrons. This work is providing the field with new tools for more critically addressing the structure of nuclei and the properties of bulk nuclear matter.

An area that at present is generating intense interest is related to nuclear processes in the cosmos. Experiments measuring neutrinos from the Sun and from cosmic-ray interactions in Earth's atmosphere strongly suggest that neutrinos are massive, a result that would imply new physics beyond the current "Standard Model" of particle physics. U.S. nuclear physicists, who have worked in the field since initiating the first experiment more than 30 years ago, are currently partners in the Sudbury Neutrino Observatory, the first detector that will distinguish solar neutrinos of different types, or "flavors." Such experiments are part of a larger effort to carefully test the Standard Model at low energies. The nucleus is a powerful laboratory for probing many of the fundamental symmetries of nature, because it can magnify subtle effects that may hide beyond the direct reach of the world's most energetic accelerators.

Another frontier area is the study of how the nucleus changes when subjected to extreme conditions, such as very rapid rotation or severe imbalances between the numbers of neutrons versus protons. Exotic nuclei play essential roles in the evolution of our galaxy: the "parents" of about half of the heavy elements are very neutron-rich nuclei, believed to have been created within the spectacular stellar explosions known as supernovae, at temperatures in excess of a billion degrees. Remarkable advances in accelerator technology have now provided the tools needed to produce such unusual nuclei in the laboratory, opening the door to new experiments on the properties of nuclear matter near the limits of binding.

The recommendations by this committee should be considered in the context of the careful planning in the nuclear physics community summarized by the Long Range Plans developed by the Nuclear Science Advisory Committee (NSAC). NSAC advises the two principal funding agencies for this field, the Department of Energy and the National Science Foundation. The Division of Nuclear Physics of the American Physical Society also played an important role, joining with NSAC to organize various town meetings for the purpose of gathering input from the community. The NSAC Long Range Plans have been prepared at about 6-year intervals (1979, 1983, 1989, and 1996). They have been influential in expressing new priorities of the field and in justifying new initiatives.[1] The 1979 and 1983 Long Range Plans, for example, identified CEBAF and RHIC as the most promising new initiatives for decisively advancing the scientific frontiers of the field. The recent adoption of a similar planning process by the European nuclear physics community is an indication of the perceived effectiveness of the Long Range Plans.

In parallel with CEBAF and the construction of RHIC, the NSAC Long Range Plans have also identified and recommended several smaller targets of opportunity. Among those currently being implemented with agency funding are an upgrade to the capabilities for producing energetic beams of short-lived nuclei at Michigan State University, the construction of new detectors for studying solar neutrinos, and the adaptation of RHIC to the investigation of previously inaccessible aspects of the proton's structure.

Both the Department of Energy and the National Science Foundation support user facilities of world-class capability and both have strong university programs. DOE supports the largest user facilities and university groups, while NSF supports user facilities at universities and many university user groups. The committee believes that the continuing programs in the two agencies are essential to the field, with the DOE emphasis on national laboratory facilities and the NSF emphasis at the universities providing complementary strengths and opportunities.

Because there exists a tradition of successful deliberation and planning within the nuclear physics community, the Committee on Nuclear Physics chose to emphasize the science rather than the process in the recommendations presented

---

[1]It is important to recognize that support for funding of these new opportunities was achieved through often painful priority decisions made by the community of nuclear physicists during the past decade. Other facilities, some unique and most still world-class, had to be sacrificed to pursue the scientific endeavors that were judged to be of highest priority. Major programs, such as the Bevalac relativistic heavy-ion accelerator at Berkeley and nuclear physics support for the Los Alamos Meson Physics Facility (LAMPF) were phased out, and a number of small university accelerators have been closed since 1980.

below. However, it would be remiss if it failed to bring into focus the funding stresses that now severely threaten the field.

At present it seems to be generally agreed by policymakers on all sides that the support of basic research is in the public interest, and there is considerable talk of increasing the corresponding budgets. However, the reality in nuclear physics, as in many other fields of research, is quite different. In 1996 the budget guidance provided by the DOE and NSF to help formulate the most recent Long Range Plan[2] for nuclear physics was for roughly constant manpower budgets. This goal has been undercut by the budgets of recent years. The cumulative result of a dollar-flat budget in the case of the DOE is that it now is 3 to 10 percent below the range of the guidance. In the case of NSF, there has been a larger decline, to about 15 percent below the 1996 guidance.

These decreases will curtail the utilization of new facilities and instrumentation and will jeopardize our nation's world-leading role in the field. This situation has arisen even as the efficient commissioning of CEBAF, the approaching completion of RHIC, new technical advances in the exploration of nuclei near the limits of binding, and discoveries in low-energy neutrino physics have made execution of the 1996 Long Range Plan all the more urgent, requiring the level of funding given in the guidance by the agencies.

Recommendation I: Discoveries in nuclear physics—new phenomena connected with the role of quarks and gluons in the nucleus, the structure and dynamics of nuclei, the nuclear physics of the cosmos, and the limits of the Standard Model—are within reach due to our recent investments in new facilities and instrumentation. With CEBAF having started on its research program of the quark-gluon structure of matter, RHIC about to embark on the study of matter at the limits of energy density, and with other recent advances in technical capabilities, a rich scientific harvest is limited by severely constrained budgets. The committee recommends the near-term allocation of resources needed to realize these unique experimental and theoretical opportunities.

Careful laboratory measurements of nuclear reactions that take place in stars have provided the foundation for some of the field's most important achievements in understanding the nuclear bases of the cosmos, including the solar neutrino problem and the origin of the light chemical elements in the Big Bang. Beams of exotic short-lived nuclei are opening up new opportunities for measuring nuclear properties and reactions in the poorly understood regions near the limits of stability. The properties of these barely stable nuclei have direct quantitative connections to the processes that fuel the stars and create the chemical

[2] *Nuclear Science: A Long Range Plan*, Nuclear Science Advisory Committee, 1996, available from the Division of Nuclear Physics, Office of Science, DOE, and the Nuclear Science Section, Division of Physics, NSF.

elements of our world. Beams of exotic nuclei hold great promise as tools for probing new nuclear properties and for testing fundamental symmetries at low energies. These considerations provide a compelling argument for constructing a next-generation facility that will use isotope separator online (ISOL) techniques to produce high-intensity, high-resolution beams of short-lived nuclei over a broad mass range.

Recommendation II: The committee recommends the construction of a dedicated, high-intensity accelerator facility to produce beams of short-lived nuclei. Such a facility will open up a new frontier in nuclear structure near the limits of nuclear binding and will strengthen our understanding of nuclear properties relevant to explosive nucleosynthesis and other aspects of the physics governing the cosmos.

Frontier research in nuclear physics relies on both large accelerators, such as CEBAF and RHIC, and smaller facilities, where specialized low-energy measurements can be made. These smaller facilities include several university and national laboratory accelerators where weak interaction, nuclear structure, and nuclear astrophysics studies are done. Both small and large accelerators rely critically on innovative instrumentation to make new discoveries. In the case of CEBAF and RHIC, the quality of the physics programs depends on specialized detectors. The development of much of this equipment is on a scale that is suitable for university laboratories, where graduate students can participate in the construction and gain experience with cutting-edge technology. Many of the equipment needs at the smaller facilities are equally specialized. Examples include atom and ion traps designed for precision studies of weak interactions and sensitive detector arrays for measuring nuclear reactions at the very low energies characteristic of stars like our Sun.

Recommendation III: The committee recommends continued investment in instrumentation for research. As new discoveries come to light and new ideas for experiments emerge, upgrades of detector systems at CEBAF and RHIC and instrumentation needs at smaller laboratories should be considered in accordance with their potential for new discoveries. NSAC is well positioned to provide DOE and NSF appropriate advice on relative priorities and specific major upgrades.

To foretell the course of a science beyond the near term is always difficult, as it depends both on the discoveries of the next few years and the doors that new advances in technology will open. The following represents some of the future options, among a number of attractive possibilities that can be perceived at the present time, for possible implementation in the early part of the next century.

CEBAF probes nuclei at length scales where the quark and gluon substructure of nuclei should first become apparent. It thus represents a first step in probing the relationship between standard nuclear physics based on protons, neutrons, and mesons, and the underlying fundamental degrees of freedom—quarks and gluons. To understand the transition between these regimes, it may be necessary to extend the measurements to even finer resolution, such as that offered by a 15- to 30-GeV electron accelerator. The construction of a 25-GeV machine is now under discussion in Europe, and future upgrades of CEBAF are being considered in the United States.

RHIC is about to open a new door to ultrahigh energy densities in nuclear matter. The potential discovery there of a new phase of matter—a plasma of quarks and gluons—could point the way to issues requiring still higher beam intensities or energies. Construction of the Large Hadron Collider (LHC) at CERN in Europe has recently begun, with U.S. participation. Early in the next century, this facility will allow collisions of nuclei at 40 times the beam energy of RHIC. Future discoveries at RHIC will guide upgrades of RHIC and the participation of U.S. nuclear physicists in the LHC effort.

The impact of the discovery that neutrinos may have mass will be felt throughout physics. Thus, following the Sudbury Neutrino Observatory, there may be an urgent need to develop and deploy detectors capable of exploring the spectrum of lower-energy solar neutrinos, or of greatly improving the sensitivity to neutrinos from the next supernova neutrino burst. Terrestrial neutrino experiments have put important constraints on neutrino properties; a compelling case may arise for new terrestrial experiments.

Studies of fundamental symmetries in nuclei can isolate and enhance new phenomena beyond the Standard Model. In particular, new experimental searches for a neutron electric-dipole moment and precision measurements of beta-decay correlation coefficients can become the most stringent constraints on our understanding of fundamental symmetries. Promising possibilities exist for developing sources of cold and ultracold neutrons of unprecedented intensity.

Recommendation IV: Within the ten-year time frame envisioned for this report, new discoveries will provide strong arguments for one or more major new endeavors. Possible candidates include a higher-energy electron machine, capability for the study of heavy-ion collisions with increased energy densities, new detectors to explore mass effects on the solar and supernova neutrino fluxes, and an ultracold neutron facility providing an order-of-magnitude increase in the neutron densities for studies of fundamental symmetries. The committee recommends the continuation of frequent NSAC Long Range Plan efforts, to help retain the responsiveness of the field to the most promising new opportunities.

Nuclear physics not only advances the frontiers of knowledge but also makes remarkable contributions to the needs of society. The generation of nuclear energy, both for civilian power consumption and for nuclear weapons, has had a profound impact on our society in the last 50 years. Equally far-reaching has been the impact of nuclear physics in medicine; results of nuclear physics and nuclear physics techniques, from magnetic resonance to detector technologies to the use of isotopes, have led to remarkable advances in diagnostic and therapeutic power. Nuclear diagnostic techniques have a growing and pervasive role in industry, national security, nonproliferation, geophysics, global climate research, and paleontology. Nuclear physics is the basis of important technologies in the design and preparation of materials. Through such applications, through the technical and intellectual intersections of nuclear physics with other fields of science, and through its intrinsic intellectual challenges, nuclear physics stands as one of the core sciences in the continuing advancement of knowledge.

Facilities and instrumentation are essential for progress, but science ultimately depends on the people who carry it out—on their individual creativity, drive, and enterprise. The scientists who conduct experiments and develop the theoretical framework for interpreting the results are the most essential components of the field. The continued intellectual vitality of nuclear physics as a science, and the continuation of the field's more direct contributions to societal needs, depend critically on the capacity to educate the next generation of physicists. Past performance has demonstrated that students trained in solving the enormously challenging problems of forefront physics research develop the array of skills needed to lead the nation in harnessing the rapidly advancing technology that often emerges from the research itself.

The remainder of this report summarizes the current status of the science of nuclear physics. Several items of more general interest are highlighted in boxes throughout the scientific chapters.

1

Introduction

Research in nuclear physics is an integral part of the search for knowledge and understanding of the world in which we live. All matter is composed of a hierarchy of building blocks. Living creatures, as well as our inanimate surroundings, are made of molecules, which are in turn made of atoms, whose mass resides almost entirely in the nuclei. The nuclei are composed of protons and neutrons, which ultimately consist of quarks and gluons. In the recent past, as our progress in understanding has reached down to ever smaller scales, each hierarchical level has developed its own subdiscipline, with its own distinct experimental and theoretical endeavors and new insights. Each subdiscipline has produced its own range of applications, benefiting the development of society and contributing to the scientific and technological base on which our industrial and economic strength rests.

The science of nuclear physics concerns itself with the properties of "nuclear" matter. Such matter constitutes the massive centers of the atoms that account for 99.9 percent of the world we see about us. Nuclear matter is within the proton and neutron building blocks of these nuclei, and appears in bulk form in neutron stars and in the matter that arose in the Big Bang. Nuclear physicists study the structure and properties of such matter in its various forms, from the soup of quarks and gluons present at the birth of our universe to the nuclear reactions in our Sun that make life possible on Earth.

ORIGINS AND FUNDAMENTALS

Our awareness of the very existence of a heavy nucleus at the center of the atom dates from the work of Rutherford in the first decades of this century. This work was followed by basic, exciting developments: the discovery of neutrons, of nuclear reactions and the transmutations of elements, of isotopes, of the detailed nature of radioactivity. These discoveries followed in quick succession, in parallel with the developing insight that a revolutionary new framework—quantum mechanics—was needed to describe phenomena on the scales of the atom and the nucleus. This period also initiated our understanding of how nuclear processes fuel the Sun.

The early sequence of discoveries led, during World War II, to the Manhattan Project, which was based on the prior investigation of basic nuclear properties and played a key role in the history of our nation and the world. Applications of nuclear techniques to benefit human health started early, with major developments in this field continuing to the present.

The 1950s and 1960s saw the conceptual development of basic models of the atomic nucleus that provided a successful, if approximate, phenomenological framework for describing nuclear structure and reactions. The roots of many of today's urgent questions in nuclear physics can be traced to this period: Why were such model descriptions of the nucleus so successful? How do they arise as low-energy, long-distance representations of the more fundamental theory of quarks and gluons? How do the symmetries that govern the strong interactions constrain this framework and influence the different temperature regimes? What are the limitations of the models as smaller length scales and higher energies are probed? Studies of nuclear beta decay undertaken in this period helped establish the form of the weak interaction and guided the formulation of a Standard Model that has been astoundingly successful in uniting the weak and electromagnetic interactions. Yet this model remains incomplete. How can we find hints of the missing physics with low-energy precision experiments in nuclei testing the limitations of the Standard Model?

SCOPE OF THE FIELD

The fundamental questions that confront nuclear physics today have inevitably led the field to extend its horizons, both in the reach of its frontiers and in the scope of its research enterprise. The size and energy scales of present-day nuclear physics extend from the world of atomic and condensed matter physics to the more microscopic domain of high-energy physics, and at the large end of the scale, to the stars and the cosmos. It is this broad reach that makes nuclear physics so interesting to many scientists (more than 3,000 in the United States alone) and so integrally connected to other sciences. Nuclear physics both contributes to and benefits from other fields—for instance, from atomic physics for intricate table-top experiments, to high-energy physics for hall-size collider detectors. High-energy physics is concerned with the elementary particles and their interactions; it is the goal of nuclear physics to understand and explain why and

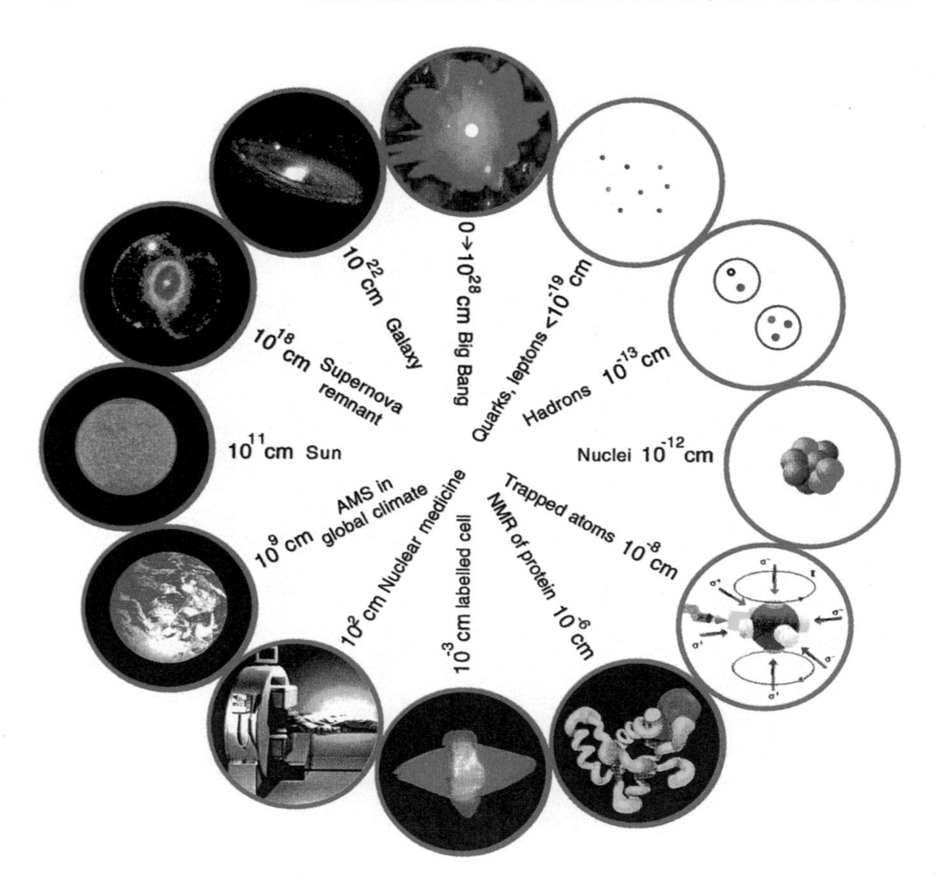

FIGURE 1.1 An illustration of the way in which nuclear physics enters into our world at different length scales. Starting at the top and going clockwise—the Big Bang, which originated from a singularity (point) in space and time; to the point-like basic constituents of matter, the quarks and leptons; to the proton and neutron building blocks in which the quarks appear in our world; to atomic nuclei; to atoms (shown isolated in an atom trap); to the structure of protein molecules determined by nuclear magnetic resonance; to labeling cells with radioisotopes; to PET scans in medicine; to studies of global climate variations through accelerator mass spectroscopy; to our Sun burning its nuclear fuel; to supernovae; to galaxies whose light shines from the nuclear reactions in its stars; and, finally, returning to the Big Bang whose remains in the present epoch encompass the universe.

how these particles, through their interactions, group themselves together to form matter.

Nuclear phenomena are important in systems spanning the entire length scale indicated in Figure 1.1, because issues requiring the application of nuclear physics methods recur at different scales with a correspondingly enormous range of energies needed for their study. All of the fundamental forces of nature come into play, each with its own special role in governing the behaviors observed over this huge span.

The central intellectual challenges around which contemporary nuclear physics research is organized drive the need for forefront accelerator facilities, sophisticated detector instrumentation, and high-powered computers, as well as the development of innovative experimental and theoretical methods. These new technologies and methods have, in turn, found rapid and widespread application, fulfilling society's needs in areas as diverse as health, security, energy production, and industrial efficiency. The application of new techniques from the physical sciences has enabled a major revolution in the life sciences, a process that will continue into the future.

RECENT ACCOMPLISHMENTS

Nuclear physics has accomplished a great deal in the recent past, as is discussed throughout this report. The committee lists here a few of the results of the past decade.

Significant advances have been made toward determining the internal structure of the building blocks of matter:

- Measurements of the distribution of quarks and antiquarks in nuclei show that these distributions are different from those in free protons or neutrons.
- Several experiments have demonstrated that neutrons and protons experience slightly, but distinctly, different nuclear forces; thus, these forces do not exactly obey the principle of charge symmetry.
- The theoretical analyses of measurements of the structure of the proton and neutron show that the spin of these particles does not have a simple origin in the quarks and that the polarization of the quark-gluon "sea" plays an important role.

New insights into the properties and structure of nuclei have been gained:

- Electron-scattering experiments have determined the structure of nuclei on a scale less than the size of individual nucleons; this advance is comparable with the first determinations of crystal structure through x-ray scattering.
- The limits of knowledge have been extended by creating in the laboratory a variety of new, short-lived elements and isotopes at the limits of nuclear stability: the first atoms of the heaviest elements through element number 112, new "magic" nuclei with unusual neutron-proton ratios, and exotic new structures such as nuclei with large, very diffuse, neutron halos.
- Superdeformed, highly elongated shapes have been discovered in nuclei

undergoing rapid rotation; the states associated with these shapes are extremely stable.

- Advanced supercomputers and new mathematical techniques allow, for the first time, exact many-body calculations of the properties of light nuclei, starting with the basic interaction between nucleons, and establishing the importance of the nuclear three-body force.

New insights have been gained in studies of matter at very high energy densities:

- Experiments have shown that systems formed in energetic collisions of nuclei decay by copious emission of sizable fragments. These decays appear to have some of the features of a phase transition from liquid-like to gas-like behavior of nuclear matter.
- Theoretical work indicates that above a certain energy density quarks will be liberated from their confinement in protons and neutrons. Recent measurements at existing accelerators have demonstrated that the relevant densities will be reached in collisions at RHIC.
- Heavy-ion collisions at very high energies have been observed to be more than simple superpositions of independent nucleon-nucleon collisions. Notably, intriguing changes are observed in the production of heavy mesons, depending on the types of quarks they contain. These changes further signal the possibilities of new phenomena in dense nuclear matter.

New results regarding fundamental symmetries have been obtained through studies at low energies:

- Precision beta-decay experiments have limited the mass of the electron neutrino to about one hundred-thousandth of that of the electron.
- The exceedingly rare process of nuclear double beta decay, by the simultaneous emission of two electrons and two neutrinos, was directly measured for the first time. Limits on double beta decay with *no* neutrinos have been greatly improved, thus testing with unprecedented sensitivity whether the neutrino might be its own antiparticle.

The connection between nuclear properties and the nature of the universe has been further elucidated:

- Precision experiments have firmly established that fewer neutrinos reach the Earth from the Sun than would be expected on the basis of solar energy production, suggesting that neutrinos have mass and can change from one kind of neutrino to another.
- Major advances have been made in measuring the cross sections for processes in stars that are crucial to the formation of the elements, utilizing both intense beams of stable nuclei and new techniques based on beams of short-lived nuclei.

- Recent theoretical progress in understanding nuclear forces suggests that the mass at which neutron stars, the densest objects in our universe, collapse into black holes, is substantially lower than previously thought. This would imply that many compact objects detected by astronomers in binary systems, and previously thought to be neutron stars, are instead likely to be black holes.
- Recent advances in the nuclear physics of supernovae have led to a deeper understanding of their explosion mechanism and of the processes responsible for heavy element synthesis, and to the discovery of a new nucleosynthesis process driven by neutrinos.

Technical innovations at a number of laboratories have made possible further advances in experimental research expanding the horizons of the research effort. In many cases, the development of these technologies generates novel applications in areas outside of nuclear physics research:

- Superconducting technology has been applied to a number of new nuclear physics accelerators: new accelerating structures for both heavy ions and electrons and new superconducting magnets for cyclotrons and beam transport.
- The CEBAF accelerator, the first continuous-beam electron accelerator at multi-GeV energies, has been operating successfully as the world's finest electron microscope for studying the physics of the nucleus.
- New techniques have been developed to produce and separate copious beams of short-lived isotopes far off stability.
- "Cooled" beams in storage rings have been developed for high-resolution studies with internal targets and used successfully in important, high-precision experiments.
- Gaseous polarized targets, in which the nuclei of atoms have their spins aligned, have been developed and used in a number of fundamental experiments.

Finally, techniques and knowledge from nuclear physics have been applied to societal needs:

- Applications of new detection techniques for scanning radioisotopes in patients have expanded the horizons of diagnostic medicine.
- The use of polarized nuclear magnetic resonance of noble gases in tomography is providing enhanced images of the lung.
- Applications of accelerators to therapeutic medicine are making significant advances; treatments with beams of protons and neutrons are becoming routine.
- Accelerator mass spectroscopy is used increasingly in a number of fields beyond archaeology—for instance, in areas related to the environment, such as geochemistry, geophysics, and global climate history, and in studies of the origins of life, as well as in medical studies.

INTELLECTUAL HORIZONS

The focus of the field is on matter whose behavior is governed primarily by the strong interaction. This class of matter includes most of the known mass in the universe, ranging from the smallest stable particles with strong interactions—the proton and the neutron—through atomic nuclei, and up to neutron stars and supernovae. The goal is to understand how all of this strongly interacting matter is assembled, how the properties of and the limitations on the existing forms of matter are predetermined by the properties and interactions of its fundamental constituents. This is a formidable task, but a central one in seeking to understand our world and our universe. The major questions facing nuclear physics at the dawn of the new millennium are considered in the discussion that follows.

The strong interaction that binds nucleons together in nuclei is much more complex than the electromagnetic force that holds electrons in atoms, and atoms in molecules. Studies of scattering between two nucleons demonstrate that their low-energy interactions can be described in part in terms of the exchange of mesons, particles of medium mass. This insight is the basis for many successful models of nuclear structure. But our best present understanding is that the fundamental constituents of nuclei are quarks and gluons, whose interactions are described by quantum chromodynamics (QCD). Both nucleons and mesons are composites of quarks. In fact, the most remarkable property that follows from QCD is that individual quarks do not exist in isolation, but instead are always found bound with other quarks and antiquarks in such composite particles. This leads to some of the most fundamental questions in modern nuclear physics:

- How do the nucleon-based models of nuclear physics with interacting nucleons and mesons arise as an approximation to the quark-gluon picture of QCD?
- In probing ever-shorter distances within the nucleus, at what point must the description in terms of nucleons give way to a more fundamental one involving quarks and gluons?
- Does the nuclear environment modify the quark-gluon structure of nucleons and mesons?

A powerful new facility, CEBAF, was recently completed at the Thomas Jefferson National Accelerator Facility to allow examination of nuclei at length scales smaller than the size of the nucleon.

The classical nuclear physics models must also break down as more and more energy is crowded into the nuclear volume. At sufficiently high temperatures, the distinction between individual nucleons in a nucleus should disappear: the nucleons will melt, and their quark constituents will be free to roam over much of the nuclear volume. This is the state of nuclear matter we believe existed in the early instants of the Big Bang: only as the universe expanded and cooled from its fiery start did nucleons coalesce from a sea of quarks and gluons. It is the goal of the Relativistic Heavy Ion Collider (RHIC), a major new facility in nuclear physics, to study matter at the highest energy density and in the process recreate and study this transition.

- What are the phases of matter formed when ordinary nuclei are heated to the very high temperatures at which quarks and gluons become deconfined from the nucleons and mesons?
- What are the experimental signatures for a transition to new phases in relativistic heavy-ion collisions?
- What are the implications for the analogous epoch in the Big Bang?

A few minutes after the Big Bang, long after nucleons had coalesced out of the early quark-gluon plasma, the universe cooled sufficiently to allow nucleons to condense into the first light nuclei. Since that time, nucleosynthesis has continued in the centers of stars such as our Sun—it is the fusion of light elements into heavier ones that is the source of solar and stellar energy. These new nuclei, expelled into space in violent stellar explosions or by stellar winds, form the raw material for the formation of new stars and new planets. Thus, nuclear physics processes are the source of the rich chemical abundance of the world that we see about us. Many of the heavy elements on Earth are the progeny of exotic, very short-lived, neutron-rich nuclei that previously could only exist in nature in violent astronomical environments, such as supernovae. Today's nuclear physics researchers have developed the technology to produce and study some of these exotic nuclei in the laboratory:

- What is the quantitative origin of the chemical elements in the Big Bang and continuing to the supernovae we observe in our galaxy and elsewhere?
- What is the influence on element production of the properties of exotic nuclei, especially those near the limits of nuclear stability that become accessible with the advent of intense radioactive beams?
- What qualitatively new features appear in this hitherto unexplored regime of nuclei, and how do they influence stellar properties?

While it is believed that nuclei can ultimately be described in terms of QCD, more empirical models of nuclear physics have provided a realistic framework for understanding a rich array of observed nuclear phenomena. These include shell structure, which makes some nuclei much more tightly bound than others; collective rotations and vibrations of many nucleons in the nucleus; transitions between regular and chaotic behavior in nuclear spectra; and weakly bound halo nuclei with an enormous increase in nuclear size. As in any physical system, pushing nuclei to their limits reveals new features and leads to new insights and understanding.

- How do nuclei behave when pushed to the limits of their excitation energy, angular momentum, and nuclear binding?
- Can the apparently simple phenomenological models, which describe several nuclear properties so successfully, be related transparently to the basic interactions of the nuclear building blocks?
- Why do these simple models work so well?

In the framework provided by empirical models, nuclear physics has much in common with other subfields, such as condensed matter physics, where understanding the effective degrees of freedom provides the essential insight into the behavior of many-particle systems. The theoretical techniques used in nuclear physics—shell structure, collective coordinates, clustering and pairing, Monte Carlo and other large-scale numerical methods—are shared by many other subfields.

- How can the symbiosis of nuclear physics and other subfields be exploited to advance understanding of all many-body systems?

Early studies of beta decay in nuclear physics helped to provide the experimental foundation for the Standard Model of fundamental interactions. In particular, they pointed toward the existence of neutrinos, elusive particles with little, if any, mass, that interact only via the weak force. The feebleness of their interactions lets them escape unscathed from the interior of the Sun and makes neutrinos an excellent probe of the nuclear reactions that fuel the Sun deep in its interior. But several experiments have found far fewer solar neutrinos reaching the Earth than expected. Today, 30 years of effort in solar neutrino physics is culminating in the first detectors that can distinguish the different kinds of neutrino interactions, and thereby test the most far-ranging explanations suggested for the low flux. Experiments with these detectors over the next few years could well prove that neutrinos have mass.

- What is the reason for the low flux of solar neutrinos?
- Will the resolution of this problem demonstrate conclusively that neutrinos have new properties, such as a nonzero mass?
- What can studies of neutrinos from supernovae reveal about the properties of neutrino "families"?
- How will such studies help us understand stellar evolution, including the mechanism responsible for supernova explosions?

It is quite possible that the "new physics" lying beyond the Standard Model may reside at energies well beyond the reach of direct accelerator experiments. Thus, our only opportunity to find this new physics may be through the "fingerprints" it has left on our low-energy world: subtle violations of the symmetries of the Standard Model. By exploiting the rich variety of nuclear species, nuclear physicists have developed an arsenal of precision techniques in searching for massive neutrinos, tiny changes in particle interactions when time's arrow is reversed, and other phenomena inconsistent with the Standard Model.

- What are the low-energy manifestations of physics beyond the Standard Model? How can precision experiments in nuclear physics reveal them?

INTERNATIONAL ASPECTS

Since the Second World War, the United States has played a world leadership role in nuclear physics. However, the contributions of other countries, particularly in Europe and Japan, have gradually increased and are now at least on an equal footing in most areas, and superior in some. This, of course, is as it should be. Science is an international undertaking, and societies with strong economies help lead the way in the pursuit of knowledge, both as an obligation and because this pursuit is in their long-term interest. Much scientific work is done in collaboration or in a spirit of friendly competition; the scientific process thrives in an atmosphere in which the implications of important new or surprising results can quickly be checked and extended by others. Unique facilities built by many nations are accessible to researchers from around the world. U.S. researchers can be found at work in laboratories of many countries—Germany, France, Finland, Japan, Canada, and Russia, to name a few. In turn, scientists from all parts of the world are carrying out their research at U.S. laboratories, both large and small. The fabric of science is intimately tied to worldwide cooperation, sometimes by formal agreements, but most often as a matter of course by informal arrangements.

EDUCATIONAL ASPECTS

A critical component of doing scientific research is the education of new scientists. A continuing flow of young scientists in training imparts a strong vitality to the process of questioning and searching for knowledge. The majority of scientists trained in nuclear physics go on to positions in industry, business, or government. Their knowledge, familiarity with modern technologies, and problem-solving skills provide excellent preparation for the requirements of our increasingly complex, technological society. The impact of the field on undergraduate and high school education, in providing students with an opportunity for hands-on contact with scientific research and with exposure to the horizons of high technology, is a continuing contribution.

SOCIETAL APPLICATIONS

Offshoots from basic research in nuclear physics have profoundly changed our daily lives. Nuclear energy and, by extension, nuclear weapons have had a deep impact on society. The uses of radioisotopes and magnetic resonance imaging in medicine are so widespread that rarely are they associated any longer with their origins in nuclear research. In these areas and in many others, inventions of new practical applications are ongoing—new radioisotopes and radiation detectors allow positron-emission tomography (PET) imaging of the human brain's functions, neutron beams serve as bomb detectors and scan for explosives, and accelerator-based mass spectrometry permits ultrasensitive detection of trace elements.

Proton and heavy-ion beams provide effective forms of cancer therapy. Accelerators developed by nuclear physicists for basic research are used extensively in materials studies. New concepts of fission reactors may create much safer and more efficient power reactors, and heavy-ion beams may move us a step closer to an inexhaustible supply of energy through inertial fusion reactors. Even in its seemingly esoteric mission for basic research, nuclear physics trains the technical manpower that invents, implements, and operates such applications in industry, medicine, and government, including national defense. Nuclear physics is one of the cornerstones of the nation's technological edifice.

2

The Structure of the Nuclear Building Blocks

INTRODUCTION

It is a remarkable fact of nature that the average distance between protons and neutrons inside an atomic nucleus is only half again as large as the individual protons and neutrons themselves. This was first revealed by applying the same experimental technique—the scattering of electrons—to probe the distribution of electric charge within both nuclei and protons, using electron beams of higher energy to resolve the finer spatial details inside protons. Because of this close packing, the internal structure of protons and neutrons (collectively called nucleons) plays an important role in nuclear physics: it influences, and can in turn be modified by, the distribution, motion, and interactions of nucleons inside nuclear matter. This mutual influence grows in importance when, in comparison with conditions in ordinary nuclei, the nucleons are squeezed even closer together (as they are in neutron stars) or are heated to even faster motion (as in the early universe).

There are some parallels between the structures of nuclei and nucleons. Both are systems of many smaller particles undergoing strong and highly complex interactions. However, a unique feature of the structure of nucleons introduces problems unlike any faced previously by scientists studying how solids, molecules, atoms, or nuclei are constructed: the particles—quarks and gluons—that appear to be the ultimate constituents of neutrons and protons have never been found in isolation from other quarks and gluons, and thus cannot be examined in their free state. A successful theory of nucleon structure must build in interactions among the quarks and gluons that are strong enough to keep them confined within neutrons, protons, or other strongly interacting particles built from quarks and gluons. All these particles, as a class, are called hadrons.

In the theory of quantum chromodynamics (QCD), where quarks interact by exchanging gluons, confinement arises because the gluons interact with other gluons, as well as with quarks. As a result, two quarks interact more and more strongly as they are pulled further apart, as if they were connected by a rubber band. This behavior is in dramatic contrast to the rapidly weakening electromagnetic force that binds electrons to nuclei in atoms, and to the even more rapidly weakening force that binds protons and neutrons to each other in the nucleus. While the mechanism of confinement within QCD is understood qualitatively, its very strength makes a quantitative treatment extremely difficult. Obtaining a quantitative understanding of the confinement of quarks and gluons inside hadrons remains one of the greatest intellectual challenges facing physicists.

While studies of the issues surrounding confinement have traditionally been done at the interface with particle physics, nuclear physicists are playing an increasingly vital role in addressing this challenge, through experiments and theory. They seek answers to three basic questions:

- What is the structure of nucleons?
- Can QCD account quantitatively for the confinement of quarks and gluons inside hadrons?
- Is the structure of hadrons modified inside nuclear matter?

Experiments on the internal structure of nucleons will constitute a significant part of the research program at both major new U.S. nuclear physics facilities of the 1990s: CEBAF at the Thomas Jefferson National Accelerator Facility (TJNAF), commissioned in 1994, and the Relativistic Heavy Ion Collider (RHIC), scheduled for completion at Brookhaven in 1999. In addition, facilities at a number of high-energy laboratories, including the Fermi National Accelerator Laboratory (FNAL), the Stanford Linear Accelerator Center (SLAC), CERN in Europe, and the German Electron-Synchrotron Laboratory (DESY), have been and continue to be crucial for experiments probing quark and gluon distributions in nucleons and nuclei.

Major efforts in nuclear theory are devoted to developing techniques for performing at least approximate QCD calculations, and for demonstrating how the conventional treatments of nuclei—as assemblies of nucleons exchanging mesons—can be viewed as an effective low-energy limit of QCD. Successful theoretical approaches must explain the structure not only of nucleons, but of other hadrons as well.

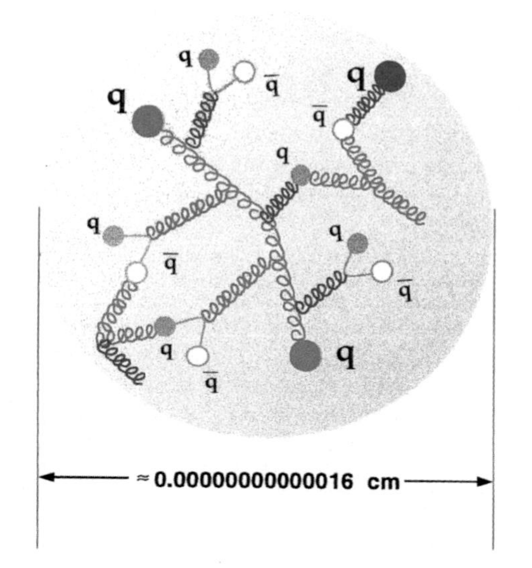

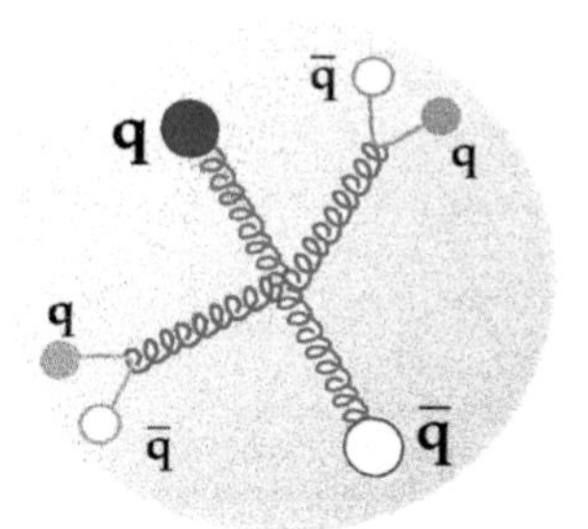

FIGURE 2.1 Schematic illustration of the substructure of a proton or neutron (left) and of a meson (right), according to the theory of quantum chromodynamics (QCD). Among the constituents confined within the nucleon are three point-like valence quarks, shown here as heavy colored dots, which interact by exchanging gluons, shown as spring-like lines. Instead of three quarks, the meson has one quark and one antiquark (dot with white center) as valence constituents. The strong interactions induce additional gluons and a "sea" of virtual quark-antiquark pairs, shown as smaller, fainter dots. Quarks are labeled "q" and antiquarks "$\bar{q}$". The colors of the constituents represent their intrinsic strong charges, the source of their participation in QCD interactions. Note that quarks appear only in groups of three (with different colors) or in quark-antiquark pairs. The nature of the strong interactions inside a nucleon and the relative contributions of various types of valence and sea quarks, as well as gluons, to the nucleon's overall properties have become major topics of research in nuclear physics.

THE INTERNAL STRUCTURE OF PROTONS AND NEUTRONS

The strength of the QCD confining interactions leads to the picture of a nucleon, illustrated in Figure 2.1, as a seething ensemble of a large and ever-changing number of constituents. A major aim of nuclear experiments through the next decade is to take detailed "snapshots" of this structure at various levels of resolution. The highest resolution is provided by highly energetic projectiles, which interact with individual quarks, antiquarks, and gluons inside a proton or neutron. These interactions, being sensitive to the motion of the struck particle, can map the probability for finding the various constituents as a function of the fraction they carry of the nucleon's overall momentum. Such detailed maps provide crucial tests for QCD calculations of the nucleon structure; a number of basic features have yet to be delineated or understood. At the same time, less-energetic projectiles must be used to obtain a lower-resolution, but more global, view of the nucleon's properties. These include its overall size and shape, distributions of charge and magnetism, and its deformability when subjected to external electric or magnetic fields. These lower-energy spatial maps serve as essential assembly drawings in understanding how the nucleon is actually put together.

First Steps

The earliest spatial maps of nucleons revealed that both protons and neutrons have sizes on the order of one ten-trillionth of a centimeter (a distance named 1 fermi, in honor of Enrico Fermi, one of the greatest of nuclear physicists). These maps were made by scattering electrons elastically from hydrogen or deuterium targets, at sufficiently high energies that the electrons could resolve structures smaller than 1 fermi. Deuterium (whose nucleus comprises only one proton and one neutron) was needed to provide a neutron target because free neutrons are radioactively unstable, decaying with a lifetime of about 15 minutes. Electron scattering probes the distribution of electric charge and magnetism within the target; the early experiments had already provided a hint that smaller, electrically charged constituents resided inside the uncharged neutron.

It took a different kind of experiment, deep inelastic scattering, to provide the first definitive evidence for the quark substructure. In this process, as illustrated in Figure 2.2, an incident high-energy electron transfers not only momentum to the target nucleon (as in elastic scattering as well) but also a large quantity of energy. The energy lost by the electron is transformed into additional particles (mostly mesons) produced in the scattering. Even though these extra particles were not detected in the early experiments, new information was gained by noting that the dependences of the scattering probability on energy transfer and on momentum transfer were strongly correlated. The observed correlation suggested that the electron had actually scattered from a very small, charged object (quark) within the proton.

Technological Advances

The techniques used in the early explorations of nucleon structure recently have been revitalized by technological advances in nuclear physics. These advances permit physicists to address more sophisticated scientific issues by allowing more detailed snapshots of the internal state of protons and neutrons. In addition, these methods have been supplemented by techniques with complementary sensitivities, using beams of high-energy hadrons.

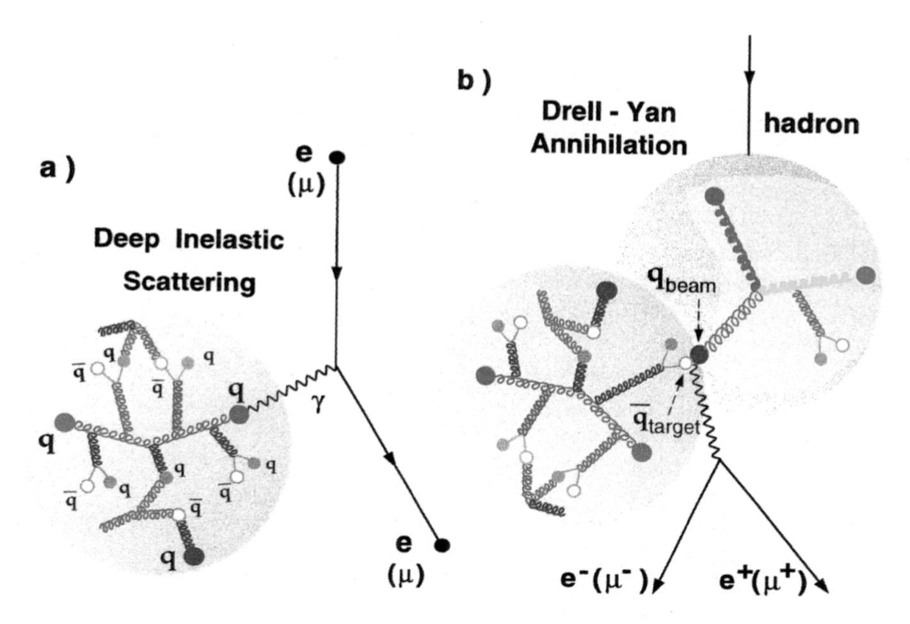

FIGURE 2.2 Two ways of probing the distribution of quarks and antiquarks inside nucleons. In (a), an electron (or a muon, a heavier cousin of the electron) scatters from a single quark (or antiquark) and transfers a large fraction of its energy and momentum to the quark via an intermediary photon (shown as a squiggly line). Such deep-inelastic-scattering measurements have provided important general information on how a proton's overall momentum is shared among its quarks. The technique has recently advanced to allow the separation of contributions from different types of quarks—for example, from strange quarks in the induced sea around the valence quarks.

In (b) (the Drell-Yan process), the induced antiquarks in the sea of the target nucleon are probed more directly when a quark inside an incident hadron has enough energy to annihilate with one of these antiquarks. The energy released by annihilation produces a "virtual" photon, which then materializes as an electron-positron pair at very high relative energy (or as an analogous pair of muons). By detecting this pair, one can probe the antiquark and quark structure of the hadrons.

Nuclear physicists have extended both these techniques to study quarks and antiquarks inside nuclei, as well as nucleons, and to probe the sharing of the nucleon's overall spin about its direction of motion, as well as of its momentum, among the quarks and antiquarks.

Among the most dramatic technical improvements relevant to experiments in this field are:

- advances in the production of spin-polarized beams and targets, and
- new electron accelerators and detector systems that allow the many particles produced in scattering processes to be detected in time coincidence.

Spin polarization refers to the preferential orientation, along some chosen direction in space, of the intrinsic spins of the particles that make up a beam or a target. With techniques now available, the spins can be aligned in the same direction for as many as 90 percent of the electrons in a beam or the protons in a target. Furthermore, polarized neutrons can be prepared inside polarized deuterium or ^{3}He nuclei, for use as targets, as indicated in Figure 2.3. The polarization is a powerful tool; it allows one to separate the effects of small but important processes that depend strongly on the polarization direction of the beam and/or target from more dominant contributions to the scattering that depend differently, or not at all, on the polarizations. For example, polarization measurements allow, for the first time, high-precision maps of the small but nonzero electric-charge distribution within the uncharged neutron, without being overwhelmed by the effects of the much stronger interactions of the electron beam with the magnetism of the neutron or with the charge of the partner proton inside a deuterium target nucleus. The new technology also provides access to new structural features, such as the spin substructure of the nucleon: deep-inelastic-scattering experiments carried out with polarized beams and polarized targets have begun to reveal how the spins and orbital motions of quarks and gluons combine to produce the overall angular momentum of a proton or a neutron.

One of the inherent difficulties in obtaining new information on hadron structure is that high energies are required to probe small objects. At these high energies, many new particles can be produced in reactions with nucleons. New structure properties can be determined if these particles can be analyzed and identified as coming from the interaction of a single beam particle with a single target particle. Thus, detectors that can track many particles simultaneously, over broad ranges of angle and energy, are required. Suitable systems, such as that pictured in Figure 2.4, are being commissioned and shortly will allow a new class of electron-scattering experiments to be performed. It is also crucial to use accelerators that deliver the electron beam continuously to the target, rather than in short time bursts, in order to attain practical counting rates for the events of interest while minimizing the rate of "accidental" coincidences among particles arising from two or more unrelated interactions in the target. CEBAF—the most powerful continuous-beam electron accelerator ever built—provides a microscope of unprecedented capabilities for studies of nucleon structure.

Electromagnetic interactions, such as deep inelastic scattering, are sensitive to the charge and magnetism of quarks, but not to other quark properties, and not sensitive at all to the uncharged gluons that bind the quarks together. To obtain a more complete picture of the nucleon substructure, it is important to complement these sensitivities with measurements of quite different processes induced by beams of high-energy hadrons. One example, illustrated in Figure 2.2, is known as Drell-Yan annihilation, where a quark from a hadron projectile undergoes matter-antimatter annihilation with an antiquark from the target (or vice versa). Studies of this process have allowed separation of structure features

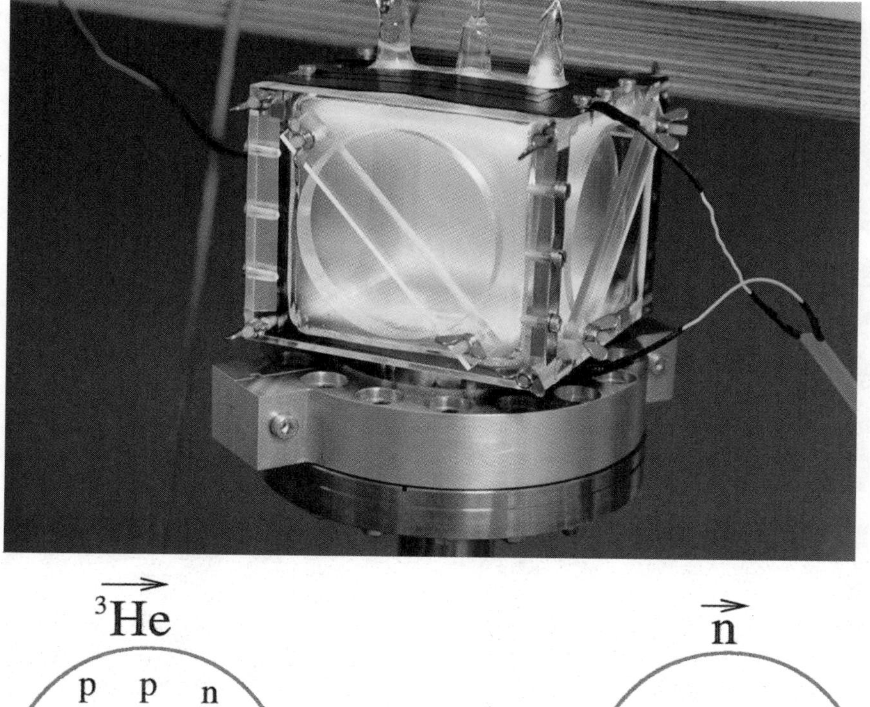

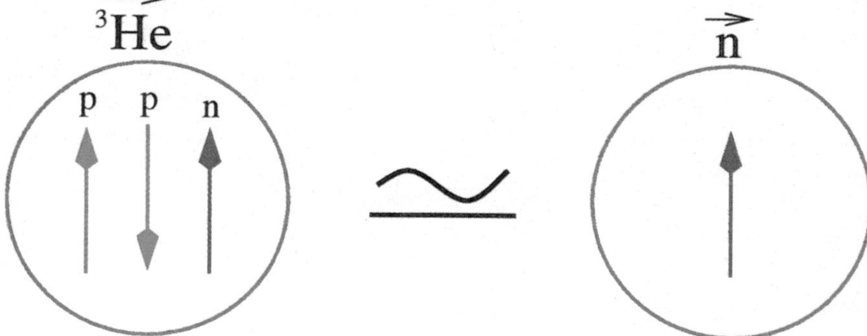

FIGURE 2.3 Among recent technological advances that facilitate dramatic improvements in measurements of neutron substructure is the development of spin-polarized ^3He targets of usable density. The photograph shows a cell filled with ^3He gas, which is exposed to laser light of appropriate frequency to "pump" most of the gas atoms into a state with the desired orientation of the nuclear spin. The purple glow is from light emitted by excited ^3He atoms. This particular target has been used in a deep inelastic scattering experiment to measure the contributions of quarks and antiquarks to the overall spin of a neutron.

The lower frame illustrates why a spin-polarized ^3He nucleus can serve as an effective surrogate for a target neutron of known spin orientation. The nucleus comprises two protons, which spend nearly all of their time with opposite spin orientations, plus a single neutron, which nearly always shares the spin preference with which the ^3He nucleus is prepared. To good approximation, it is only the quarks and antiquarks inside the neutron that contribute to changes in the scattering probability when the ^3He spin is reversed.

FIGURE 2.4 Parts of the CEBAF Large Acceptance Spectrometer. This system allows one to detect, identify, and measure the momenta of multiple particles resulting from the interaction of energetic electron beams with nucleons and nuclei. At its core is a novel superconducting magnet that produces a toroidal field capable of bending the trajectories of the emerging particles without perturbing a spin-polarized target mounted along the spectrometer's axis. Three (out of a total of six) "slices" of the superconducting coils for this magnet are shown in the photograph on the left. Detectors fit in between and beyond the coils, including the inner "drift chambers" shown on the right, which measure precisely the position where charged particle tracks cross them. When coupled with the continuous beam available from the accelerator, this spectrometer will facilitate experiments of unprecedented thoroughness to probe the structure of nucleons and their many excited states. (Courtesy of the Thomas Jefferson National Accelerator Facility.)

associated with the primary (valence) quarks versus the induced "sea" of quark-antiquark pairs. Other processes can produce gluon maps of very fast colliding nucleons, by measuring the probability to create new particles, or high-energy gamma rays, in whose formation the gluons in one or both participants play the decisive role. Experiments using such hadron probes will benefit greatly from recent technological advances. For example, at RHIC the contributions of gluons and sea quarks to the proton's spin will be discerned from the results when two very high-energy beams of spin-polarized protons collide.

Experimental Opportunities

According to the Standard Model of particle physics, quarks and antiquarks come in six different varieties, or "flavors," characterized by different masses. Only the two lightest varieties, called up quarks and down quarks, participate as valence quarks in nucleons. Indeed, in the original, simplified picture of a proton's substructure, all its properties were thought to arise from combining two up quarks and one down quark in a state with no relative orbital motion. Although this picture provides a deceptively successful account for some properties, such as the overall magnetism of the proton and neutron, it is clearly an incomplete representation of the complex quark-gluon dynamics. Experiments planned for the coming decade should yield a far more realistic picture of what goes on inside the nucleon, including the roles of sea quarks of various flavors, of gluons, and of the orbital motion of all these constituents.

The induced sea of quarks and antiquarks inside a nucleon is a fascinating manifestation of the strong confining forces that nuclear physicists seek to understand. Electrically charged particles, which interact via the much weaker electromagnetic force, have a small probability to liberate virtual electron-positron pairs from the surrounding vacuum. These particles are called virtual because they exist only fleetingly, soon to be swallowed up by the vacuum once again. These electron-positron pairs have subtle, but measurable, effects on the structure of atoms—effects whose study led to crucial early successes of the field theory known as quantum electrodynamics. In contrast, the analogous pairs of virtual quarks and antiquarks in a nucleon are excited with high probability and profoundly influence the nucleon's properties. It is a critical goal of experiments on nucleon structure to probe the relative probabilities for finding quarks of different flavors in the sea.

For example (see Box 2.1 for details), recent measurements of both deep inelastic scattering and Drell-Yan annihilation processes have refuted the naive expectation that anti-up and anti-down quarks might be found with equal probability in the proton's sea. A most intriguing implication of this discovery is that the essence of the sea may be captured in a picture where the virtual quark-antiquark pairs are thought of as mesons, which they resemble (as can be seen in Figure 2.1). In particular, the observed preference for anti-down quarks is consis-

BOX 2.1 Sampling the Proton's Flavors

How, exactly, are protons and neutrons made? Nature has guarded her secret recipes as vigilantly as the Coca-Cola Company, even to the extent of forbidding the ingredients from ever being used alone. Nonetheless, nuclear physicists are uncovering the proton's secrets little by little, exploiting ever more selective techniques to isolate individual "flavors" in the complex mélange.

Physicists have discovered six flavors, or varieties, of quarks—the tiny constituents of protons, neutrons, and all their heavier cousins. The different flavors have been given the metaphorically confused names: up, down, strange, charmed, bottom, and top. Until recently, only the two lightest types, up and down, were known with certainty to reside in a proton. Indeed, many of the proton's distinguishing features can be understood as emerging from a combination of three primary—two up and one down—quarks. Surprisingly, though, physicists believe that these primary (or valence) quarks account for only a small fraction of the proton's mass. Some of the remaining mass is associated with a "sea" of quark-antiquark pairs induced by the presence of the valence quarks. Though individual pairs appear and vanish, they are found with high probability, much as the flickering lights in a dense swarm of fireflies. The transient quarks and antiquarks in this sea can be of the up or down variety or, in principle, of any other flavor. The flavor composition of the quark sea, which gives the proton and the neutron much of their complexity and richness, is the focus of a number of ongoing experiments.

Physicists' initial naive expectation was that nature would mix different flavors in the sea in proportions depending only on the quark mass associated with each flavor. Since up and down quarks have nearly equal masses, they should then appear with nearly equal probability in both protons and neutrons. That expectation is refuted by recent results, shown in Figure 2.1.1, from an experiment at the Fermi National Accelerator Laboratory. These results imply a strong preference for down over up antiquarks in the proton's sea.

Why should nature prefer such a flavor imbalance? One possible explanation is that the proton spends some of its time as a neutron surrounded by a positively charged pi meson. But this description uses the language of conventional nuclear theory, dealing with protons, neutrons and mesons, rather than with quarks directly. One of the primary goals of nuclear physics is to understand the relations between these two languages, and the conditions under which one or the other provides the more efficient description of matter. The result of the Fermilab experiment suggests that the theoretical framework that works best for treating the structure of nuclei may also be unexpectedly useful in treating the internal structure of protons and neutrons.

Quark flavors other than up or down can appear in a proton *only* as part of the sea. At present, there is particular interest in the strangeness flavor, which corresponds to the next lightest quark mass after up and down quarks. Hints from several different experiments have suggested that strange quarks and antiquarks have a greater role in the proton than would be expected on the basis of their mass alone. A series of ongoing and planned experiments at several laboratories (an example is shown in Figure 2.1.2) seeks to distinguish strange quarks by their contribution to the weak interaction between electrons and protons. This interaction will be unmasked from the normally dominant electromagnetic interaction by searching for a characteristic violation of mirror symmetry. The violation can be

BOX 2.1 Continued

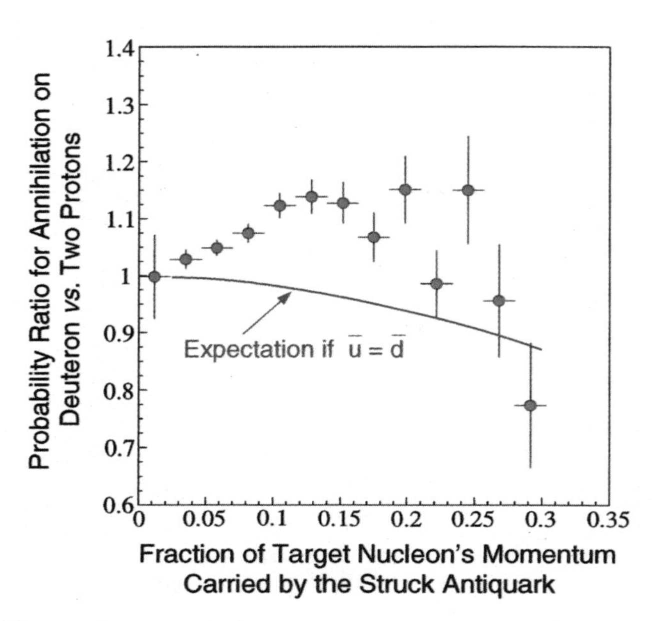

FIGURE 2.1.1 Recent results from an experiment carried out at Fermilab refute the simple expectation (blue curve) that the induced sea of quark-antiquark pairs in a proton or neutron contains up and down quark flavors in equal amounts. An experimental team led by nuclear physicists from Los Alamos National Laboratory has found the down flavor strongly prevalent in the proton's sea. They studied processes where a valence quark in a high-energy beam proton preferentially annihilates a sea antiquark in either a proton or deuteron target. A deuteron contains one neutron and one proton.

determined in precise measurements of a tiny expected difference in the scattering rates when the electron beam is prepared in two different ways: with most of the electrons spinning clockwise versus counterclockwise, as seen from the electron's direction of motion.

Many more questions remain to be addressed to discern nature's complete recipes for protons and neutrons. They concern, for example, the presence of

tent with the effect of the proton occasionally emitting, then reabsorbing, a virtual pi meson. The conventional view of nuclei, in terms of nucleons and mesons rather than quarks and gluons, may thus even provide a relevant approximation to QCD in treatments of nucleon structure. It will be interesting to see if such treatments also remain useful in addressing the presence of heavier quarks and antiquarks, such as those with the "strangeness" flavor, in the nucleon's sea. As discussed in Box 2.1, a series of experiments at CEBAF and other accelerators is

FIGURE 2.1.2 A part of the detector constructed at the University of Illinois, which is currently in use at the Bates accelerator to study small violations of mirror symmetry in electron-proton scattering. One of the collaborators is seen among the large reflectors used to focus radiation emitted by the scattered electrons. The experiment, initiated at the California Institute of Technology, is sensitive to the presence of the strangeness flavor in the quark-antiquark sea within a proton. (Courtesy of the MIT Bates Linear Accelerator Center.)

even heavier quark flavors, the contributions of different flavors to the overall spin of a proton, and possible changes in the flavor makeup between protons and neutrons, or between free protons and those embedded in nuclei. The ultimate goal is to understand how nature's choice of recipes has determined the character of the world around us.

aimed at determining the contributions of strange sea quarks to nucleon properties.

A particular nucleon property in which strange quarks may play a significant role is the spin substructure. The intrinsic spin and the associated magnetism of a proton are among its defining characteristics. They are, for example, essential in accounting for radiation observed from the heavens and in allowing magnetic resonance imaging. Measuring and understanding the relative contributions to the proton's spin from its various constituents is a major goal of nuclear physicists. The first information has been provided by recent, polarized, deep-inelastic-scattering experiments. The results, shown in Figure 2.5, suggest that the intrinsic spins of all the quarks and antiquarks combined account for only a fraction (less than about 30 percent) of the nucleon's overall spin. In particular, the net contribution from up and down quarks seems to be partially canceled by that from the strange quarks, which appear to line up opposite to the overall spin direction. However, this interpretation relies on arguments based on quite different measured properties of hadrons, and it must be confirmed by future experiments that can separate the spin contributions from valence and sea quarks, and from quarks of different flavors.

The next generation of experiments on the spin substructure will study reactions induced by both polarized lepton (electron, positron, and muon) beams and polarized proton beams on polarized protons. The results will also provide the first information on the contribution from the spin of gluons to the proton spin, a contribution that recent theoretical work suggests may be large. Experiments with colliding beams of polarized protons at RHIC will be especially well suited to probe this gluonic role. If the quark and gluon spins together do not account for the nucleon spin, the only remaining source available to "balance the books" (as indicated in Figure 2.5) is the relative orbital motion of the quarks and gluons. Thus, early in the next century, these experiments on the spin substructure may indirectly lead us beyond our current, still rudimentary understanding of quark motion inside a nucleon.

One of the major advances of the past five years was the first measurement of the spin substructure of a polarized neutron, made possible by innovative developments in spin-polarized deuterium and ^{3}He targets. These experiments permitted a quantitative test of a rigorous QCD prediction, the Bjorken sum rule, for the net difference between the quark spin distributions in a proton and a neutron. The recent data agree with the sum rule value, demonstrating that our maturing picture of the nucleon's substructure remains consistent with QCD.

In parallel with these probes of the "trees" inside the nucleon "forest," nuclear physics experiments are improving the quality of the global view of the forest itself, by applying modern technology to study the electromagnetic interactions of nucleons at lower energies. One example, illustrated in Figure 2.6, is the vastly improved precision expected from ongoing polarization experiments mapping the neutron's electric charge distribution, which is related to the overall spatial layout of valence and sea quarks. A second example is provided by recent measurements of the distortions induced in a nucleon's charge and magnetization distributions when it is exposed to external electric and magnetic fields. These properties have been measured by scattering beams of photons—the quantum carriers of the electric and magnetic fields—from nucleons. Greater flexibility in the nature of these stimulating fields can be obtained, and hence a number of new features of the nucleon's response can be measured, in future experiments in

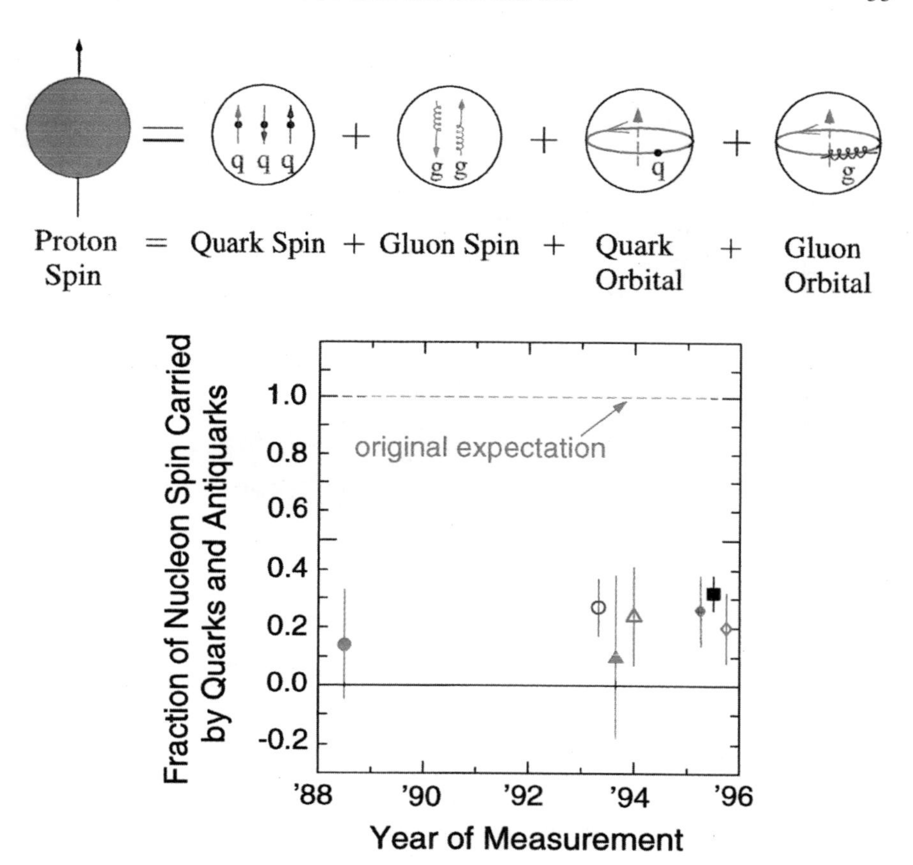

FIGURE 2.5 How do the proton's various constituents contribute to its overall spin? As illustrated by the upper diagram, the quarks, antiquarks, and gluons are all believed to have their own intrinsic spins, and these must contribute. But so also must the relative orbital motions of the quarks and gluons inside the proton. Measuring and understanding the relative contributions from these different sources is a major goal of nuclear physicists.

The first measurements of the proton's spin substructure have been made recently, employing the technique of deep inelastic scattering with spin-polarized beams bombarding spin-polarized targets. By combining these measurements with constraints from other data, one can infer the fraction of the proton's spin carried by the intrinsic spin of quarks (and antiquarks) of different flavors. The results of experiments performed at CERN, SLAC, and DESY, summarized in the graph, point to an unexpected outcome: all the quarks and antiquarks together account for no more than one-third of the total spin. More direct probes of the spin alignment of different flavors of quarks, separation of the contributions from quarks and antiquarks, and extraction of information on the gluon spin contributions are goals of ongoing and planned second-generation experiments.

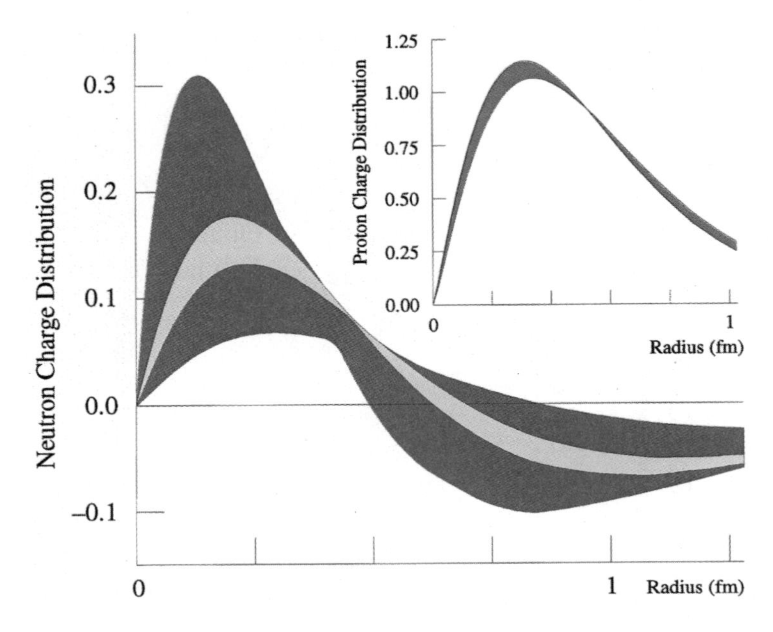

FIGURE 2.6 Although the neutron and proton are equally important in building nuclei, the quality of experimental information available about their internal structure has been vastly different, because one cannot make a target of free neutrons. For example, the graphs here compare what we know from electron-scattering experiments about the amount of electric charge per unit radial distance from the center of a neutron (on the left) versus a proton. The width of the blue band for the neutron and of the curve for the proton indicate the approximate level of uncertainty in the currently available experimental information. The crude data for the neutron suggest that smaller charged particles do indeed reside inside, with positive charges preponderant near the center and negative charges near the periphery, but provide little additional constraint on nucleon structure calculations. Our knowledge of the neutron will be brought closer to a par with the proton (as suggested by the yellow band on the left) by a new series of experiments, made possible by the advent of continuous-beam electron accelerators and by advances in the technology for measurements of spin polarization.

which the real photon beams are replaced by virtual photons, emitted and absorbed when an electron interacts inelastically with a target nucleon.

The proton and neutron represent the lowest energy states of a family of hadrons that can be constructed from up and down valence quarks taken in combinations of three. The energies of the higher-lying states, and the probability that a nucleon may be elevated into one or another of them by exposure to photon or electron beams, carry additional information about the quark substructure. The new advances in accelerators and detectors, particularly at CEBAF with the apparatus pictured in Figure 2.4, will, over the next several years, allow a dramatic increase in detailed information on the properties of these excited states. These states are referred to as resonances, rather than as particles, because they exist for only a tiny fraction of a second; one detects their excitation by looking at the particles emitted when the resonances decay or de-excite. The study of these resonances is complicated by the fact that their spectrum is no longer really discrete: their very short lifetimes imply (by the Heisenberg uncertainty principle) that each of them can be produced over a broad range of excitation energies, and the energies overlap for different resonances. Spin-polarized beams and targets will be essential in disentangling the effects of individual resonances and measuring their structure.

In summary, the coming generation of experiments will assemble a rather detailed, multifaceted empirical picture of the internal structure of protons and neutrons. If QCD is a valid theory for the confinement of quarks inside hadrons, then, eventually, one or more of the techniques developed for solving QCD (see the following subsection) must reproduce this empirical picture. In the meantime, the experimental results serve as an invaluable guide in the construction of structure models, which seek to bridge the gap between the nucleon-meson and quark-gluon descriptions of hadronic matter. The evolving empirical picture of the nucleon's substructure provides a necessary baseline against which to compare the measured properties of nucleons when they are embedded inside nuclear matter.

ACCOUNTING FOR CONFINEMENT: FROM QCD TO NUCLEAR THEORY

The best-established calculational technique for quantum field theories works when the fundamental interaction is quite weak. This is true in QCD only when two quarks have a separation very much smaller than the size of a hadron. Indeed, the theory is quite successful in this regime in accounting for hard collisions between hadrons at very high energy. However, new theoretical approaches are required for QCD to confront its most important manifestation in the matter around us: the phenomenon of quark confinement. Two approaches are being actively pursued. In one, full numerical solutions to the field theory are obtained, using the most powerful existing computers, but even this requires artificially restricting the hadron's constituents to exist at certain discrete points in space and time. The second approach concentrates on "effective" theories, which retain the crucial symmetries of QCD but relegate other details to parameters that must be adjusted to reproduce several key experimental results. These effective theories hold the promise to establish a clear connection between QCD and the more phenomenological treatments developed over decades to account for observations of nuclear structure and reactions. Measurements of hadron properties provide critical guidance in selecting appropriate crucial QCD features, in determining parameter values, and in testing the self-consistency of these calculations. In particular, experimental searches for new mesons and for new excited states of the proton and neutron have the potential to unveil predicted missing links, to reveal elusive quark-gluon properties omitted from model calculations, or to suggest unforeseen symmetries in the complex QCD interactions inside hadrons.

Working with Quarks and Gluons

The masses of atoms and nuclei are nearly equal to the sums of the masses of all their constituents; the interactions among the nucleus and the electrons in an atom, or among the many nucleons inside a nucleus, cause only slight changes in the overall mass. The situation in a hadron is radically different! The bare masses of the three valence quarks in a nucleon are believed to contribute no more than a few percent to the overall nucleon mass. Instead, the nucleon mass—and therefore the mass of our visible world—is dominated by the strong confining interactions, as manifested in the high probability for finding virtual quark-antiquark pairs or gluons inside the nucleon, in addition to the valence quarks. The strength of these interactions makes it at once essential and exceedingly difficult to calculate the effect of the interactions reliably, in order to interpret measurements of hadron properties.

The difficulty in QCD calculations of hadron structure is epitomized by the fate of a basic symmetry property of QCD, chiral symmetry. Chiral symmetry considerably simplifies the interactions of quarks in the limit in which their masses are considered as negligibly small, by forbidding gluon-exchange from reversing the "handedness" (or chirality) of quarks (i.e., from flipping one that spins clockwise when viewed along the direction of its motion, like a right-handed screw, into one spinning in the opposite direction, analogous to a left-handed screw). All of the "excess baggage" accompanying the valence quarks inside a hadron makes them appear much heavier than they really are, and thereby causes a violation of this symmetry. A major goal of approximations to QCD is to understand quantitatively the extent of this violation, together with its dependence on the density and temperature of the nuclear matter in which the hadrons are placed.

The most promising approach to approximating quark confinement and chiral-symmetry breaking in hadrons is to solve QCD numerically on a computer. The approach is approximate because it reduces the problem to one involving only a finite (and small) number of bodies. This is done artificially, by restricting each constituent to occupy one of a discrete set of points, distributed on a regularly spaced lattice in space and time. The extraction of reliable physical predictions hinges on demonstrating that the results of such calculations converge to unique answers as the lattice spacing is reduced toward zero and the lattice volume is increased. Both of these changes are severely limited in extent by available computing power. Much of the ongoing theoretical development of this approach is devoted to finding more efficient ways of formulating the QCD problem, to ease the computing crunch. Nuclear theorists are deeply involved in these developments.

Already, however, lattice QCD calculations have achieved considerable success in reproducing some basic observed properties, such as mass ratios, sizes, and decay characteristics for various hadrons. Such calculations have led to quantitative estimates of the extent of chiral-symmetry breaking, by predicting the probability of finding virtual quark-antiquark pairs and gluons inside hadrons. More importantly, they have quantified QCD predictions of two entirely new phenomena, which nuclear physicists will be subjecting to experimental tests throughout the next decade. One is a predicted phase transition, in which quarks will shed their excess baggage and chiral symmetry will be restored, in strongly interacting matter at sufficiently high temperature. The search for this new phase of matter in collisions of relativistic nuclei at RHIC is described in Chapter 4. The second prediction is for the existence of a new class of mesons, which do not have a valence quark and antiquark, as do typical mesons, but rather consist of gluons alone (gluonium). The experimental search for gluonium is described in Box 2.2. If these predictions are not confirmed by experiments, the mechanism of confinement postulated in QCD will have to be questioned.

In addition to particles composed of *pure* glue, QCD also predicts the existence of hybrid hadrons that contain more complicated valence configurations than three quarks or a single quark-antiquark pair. Discovery of the corresponding particles is an important goal of nuclear physics. Indeed, in contrast to this rich array of possibilities, nearly all of the hadrons known to date seem to be explainable in terms of the simplest quark configurations. Their measured properties can be reproduced surprisingly well if one considers the constituents to be massive pseudo-quarks, which somehow combine features of valence quarks, sea quarks, and gluons. Theoretical models based on these effective building blocks can provide qualitative insight into hadron structure that complements attempts to solve QCD numerically. But a true breakthrough in insight awaits a more detailed understanding of the relationship of the pseudo-quarks to the real fundamental particles of QCD.

Working with Nucleons and Mesons

The computational barriers are too high to imagine applying lattice QCD calculations any time soon to predict the interactions among two or more hadrons, despite their obvious importance for understanding nuclei. Indeed, these interactions appear quite complex within the framework of QCD: quarks and gluons interact by virtue of their intrinsic "color charge," but hadrons are constrained by the confinement mechanism to have no net color charge. The force between nucleons, for example, must reflect the sort of indirect effect suggested in Figure 2.7, wherein two or more momentary leakages of color between the

BOX 2.2 Where's the Glue?

Nuclear physicists study matter whose behavior is governed by the strongest of nature's forces. The smallest observed particles that feel and are capable of exerting this force are called hadrons, and an entire zoo of them has been discovered. And the search is on for even more hadrons, but now for a new, exotic species that holds the key to understanding all the rest.

The hadron zoo includes not only the protons and neutrons that build nuclei and the pi mesons that are exchanged between protons and neutrons when they interact, but also many heavier cousins of both of these types of particle. It was to introduce some order that physicists invented the concept of quarks as the building blocks of hadrons. But quarks alone were not enough. The Heisenberg uncertainty principle dictates that particles confined to a volume as small as the interior of a hadron must be moving very fast, so there must be some "glue" to hold them together. No problem: physicists invented another new particle, the gluon, whose exchange between quarks was postulated to provide this binding force, in somewhat the same way that the exchange of photons between nuclei and electrons provides the electric force that binds atoms.

Now, there is clear experimental evidence that high-energy particles interact with individual quark-like particles inside a proton. But the fact that no one has ever succeeded in isolating a *free* quark or gluon adds intrigue to the story. This fact suggests a binding force so strong that it is effectively impossible to overcome. The remarkable success of quantum chromodynamics (QCD) is that it is the only viable field theory proposed for the strong interaction in which this confinement of quarks and gluons within hadrons arises naturally. The important remaining question is this: Is QCD the correct theory of hadron structure? What testable predictions does it make that arise inevitably from its mechanism for confinement?

Confinement arises in QCD from a simple, but radical, postulate: the gluons themselves carry a strong interaction charge, in contrast to photons, which carry no electric charge. QCD then predicts that gluons can interact among themselves, and can even assemble to form exotic quarkless mesons. Such hadrons, referred to as glueballs or gluonium, *must* exist if the confinement mechanism in QCD correctly describes reality. And so, the search for them is on, with high priority.

How does one try experimentally to distinguish a glueball from conventional mesons, whose primary constituents are a single quark plus a single antiquark? The glueballs must be electrically neutral, and they may take on masses unexpected for conventional mesons, or quantum properties forbidden to them. Since gluons can produce quark-antiquark pairs, the glueballs can decay to conventional mesons, but they should do so "democratically," with little regard for the flavor of the daughter mesons they spawn. And glueballs should be produced preferentially in relatively quark-free environments, such as the aftermath of matter-antimatter annihilation.

Sophisticated detectors (Figure 2.2.1) have been built to sift through the debris of collisions where the quarks in a proton and the corresponding antiquarks in an antiproton extinguish one another, in search of decay daughters or granddaughters of potential glueballs. Elaborate sleuthing methods have been developed for the necessary multistep reconstructions of the "scene of the decay." They involve energy-sharing diagrams, such as the colorful one shown in Figure 2.2.2), where the ultimate progenitors appear as enhanced lines at characteristic energies.

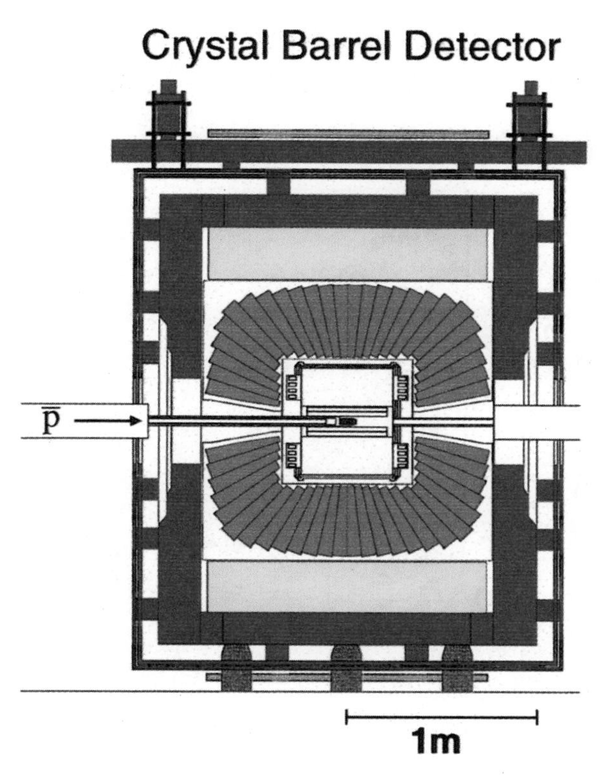

FIGURE 2.2.1 Cross-sectional side view of the detector used to search for new mesons in the Crystal Barrel experiment carried out at CERN. The detector sifts through the quark-poor debris of proton-antiproton annihilation collisions, in search of charged particles and high-energy gamma rays that may result from the decay of previously undiscovered particles. The experiment takes its name from the cylindrical array of 1380 crystals (colored here in cyan), used to measure gamma-ray energies with good resolution. (Courtesy the Crystal Barrel experiment group.)

The data in the energy-sharing diagram here reveal the existence of a previously unknown meson, with properties strongly suggestive of the predicted lightest glueball. However, its not quite "democratic" decay probabilities suggest a possible Jekyll and Hyde particle, which spends part of its time as a glueball and the other part as a conventional meson. A more convincing demonstration of the presence of the glue awaits future experiments, intended to unveil other members of the glueball family, including pure ones, whose quantum properties forbid hiding in conventional clothing. Such searches are painstaking, needle-in-a-haystack affairs, but they hold promise of a dramatic payoff—discovery of one of the few basic missing links in physicists' overall scheme for the structure of matter.

Box continued on next page

BOX 2.2 Continued

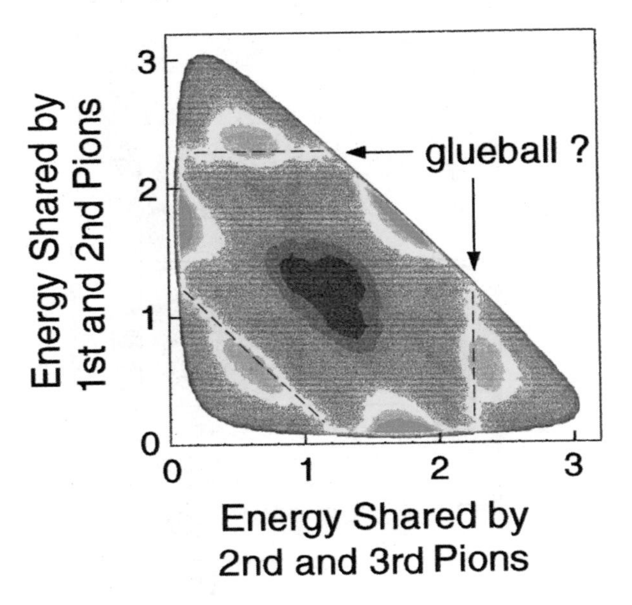

FIGURE 2.2.2 Data from the Crystal Barrel experiment reveal the existence of a previously undiscovered meson with mass and quantum properties near those predicted by QCD for the lightest particle made from gluons alone. The different colors indicate the relative probabilities for different ways of sharing the total energy from the proton-antiproton annihilation among three detected pi mesons. The straight yellow bands, with dashed lines superimposed, are populated by events where two of the three pions result from the decay of the newly discovered meson. (Courtesy the Crystal Barrel experiment group.)

nucleons compensate for one another. How can this view be joined with the different picture that has so successfully accounted for the scattering of nucleons and for nuclear structure at low energies? In that picture, the strong interaction between low-energy nucleons is mediated primarily by the exchange of pi mesons (called pions for short) between them. A crucial step along the path from quarks and gluons to nuclei—a path being scouted by many nuclear theorists—is to understand why the pion is "chosen" as such a dominant manifestation of QCD interactions at low energies.

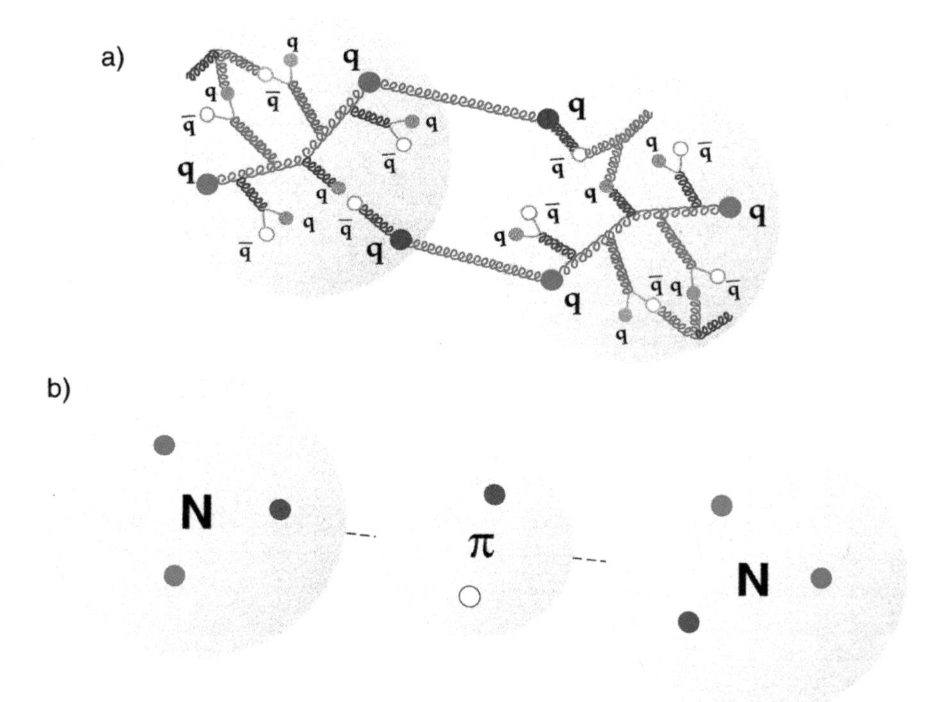

FIGURE 2.7 The simplest contributions to the force between nucleons, as viewed from (a) QCD and (b) conventional nuclear theory. In (a), the exchange of two colored gluons causes two quarks in each nucleon to change their colors (blue changes to green and vice versa in the case illustrated). This process produces a force without violating the overall color neutrality of the nucleons. The strength of the force depends on the separation of the different quark colors within each nucleon. On the other hand, low-energy nuclear physics measurements show clearly that the longest-range part of the force arises from the exchange of a single pi meson between two nucleons, as in (b). In this low-energy view, the internal structure of each nucleon is generally attributed to three pseudo-quarks, which somehow combine the properties of the valence quarks, sea quarks, and gluons predicted by QCD. Nuclear physicists are seeking to understand how these and other contributions combine to give the overall nuclear force.

The special role of the pion within QCD is that of primary agent for the breaking of chiral symmetry in nucleons and nuclei. An up or down quark can change from left-handed to right-handed (or vice versa), violating chiral symmetry, if it emits or absorbs a pion rather than a gluon. If the pion had zero mass—a so-called Goldstone boson—the symmetry would still be preserved in nuclear interactions as long as pions as well as quarks were included within the system considered. It is thus the pion's unusually low mass, in comparison with all other mesons, that keeps the symmetry at least approximately satisfied in nature. As near-Goldstone bosons, pions interact with one another, and with nucleons, fairly weakly when they have low momentum. This observation suggests that the more traditional methods for solving quantum field theories when the interactions are weak may be applicable to QCD at low energies, if the theory can be expressed in terms of nucleons and pions as the main players, rather than quarks and gluons. Developing and testing such an effective field theory, known as chiral perturbation theory, is an important theme in contemporary nuclear physics. The forms of the allowed interactions among pions and nucleons are restricted by the symmetries of QCD, but experiments must be used to determine their precise strengths, which are still too hard to calculate directly. Once the strengths are fixed, chiral perturbation theory can predict as yet unmeasured properties of pion interactions.

Experimental tests of these predictions, and hence of the self-consistency of the effective theory, evaluate the ability of QCD to deal with chiral-symmetry breaking at low energies. Important ongoing tests involve the study of decay modes of heavier mesons to pions and photons, the probability of producing pions in photon-proton collisions, and the interaction of two pions when they have essentially no relative momentum. The most unambiguous way to measure this last interaction calls for forming an exotic cousin of the hydrogen atom in the laboratory, one in which the proton is replaced by a positively charged pion and the electron by a negative one, orbiting each other under the influence of their mutual electric attraction. Experimental attempts to study this system combine atomic, nuclear, and particle physics techniques to shed light on QCD and its connections to conventional nuclear theory.

HADRONS IN NUCLEAR MATTER

Nucleons and mesons are the building blocks of ordinary nuclear matter, but there is no guarantee that these building blocks have properties in nuclei identical to those of the isolated hadrons we study in the laboratory. According to QCD, the properties of hadrons are strongly influenced by the induced sea of quark-antiquark pairs and by the gluons produced in the confining interactions. But the conditions for achieving the most favorable overall energy balance, and hence the probability for exciting these virtual particles, may well be different in a group of closely spaced nucleons than inside an isolated nucleon. Indeed, there are theoretical predictions that the probability of finding virtual quark-antiquark pairs decreases systematically (thereby partially restoring the chiral symmetry of QCD) as the density of surrounding nuclear matter is increased; at very high densities this probability vanishes. If such changes can be definitively observed in experiments planned for the coming decade, it will have a profound influence on our understanding of quark confinement, of nuclear matter, and of the densest stars.

If the quark-antiquark pair probability decreases inside nuclear matter, so should the masses of most hadrons, and their sizes and interactions may change correspondingly. It is an important goal of future experiments to search for direct evidence of such modifications of hadron properties. Several techniques have been applied in searches to date. Measurements sensitive to the internal spatial structure of nucleons have been made by scattering beams of electrons and neutrinos from individual nucleons inside nuclei. Attempts have been made to determine the masses of mesons produced in nuclear matter by studying their decay, especially to such daughter products as leptons (electrons, positrons, and muons), which are not themselves significantly affected by passage through the surrounding matter. No generally accepted evidence has yet been observed for significant modifications at normal nuclear densities. Indirect hints of appreciable changes are afforded by systematic differences observed in the spin-dependent interactions of nucleons with other nucleons inside versus outside nuclei. The best hope for definitive tests of the predicted variations may lie with measurements of the properties of mesons produced in rather high-density matter formed in the collisions of energetic heavy nuclei.

A complementary approach to searching for changes in nucleon structure inside nuclear matter is to probe the distribution of quarks and gluons in nuclei, using high-energy beams of leptons or hadrons, via the deep-inelastic-scattering and Drell-Yan annihilation processes. When the nucleus is viewed under such a powerful microscope, one loses sight of specific hadrons; the struck quark or gluon may reside in any of the nucleons or mesons present in the nucleus. The deep-inelastic-scattering measurements shown in Figure 2.8 have already demonstrated that the quarks share the overall momentum of a nucleus in a manner clearly, but mildly, different from what one would expect for an ensemble of free nucleons. The interpretation of these differences must take account of relatively mundane effects, such as the binding of the nucleons and the presence of virtual mesons being exchanged between them, along with any systematic effect of nuclear matter on the internal structure of the nucleons. Calculations suggest that the mundane effects are not sufficient to account for all of the deviations seen in Figure 2.8. The residual effects are consistent, for example, with models in which the size of a nucleon changes slightly, but systematically, in nuclear matter.

The interpretation of the deep-inelastic-scattering experiments is aided by comparison to results from Drell-Yan annihilation, which, unlike the former process, can be used selectively to probe the antiquarks in the nucleus, yielding greater sensitivity to the presence of mesons. Surprisingly, Drell-Yan data show no evidence of the surplus of antiquarks predicted to arise from pions being exchanged between nucleons. This result is confirmed by nuclear reaction measurements, which have searched for the characteristic change in a proton's spin orientation when it absorbs a pion from the nucleus. The failure to observe the expected pion excess—a basic manifestation of nuclear forces—may itself reflect changes in hadron properties (e.g., in the size of nucleons) inside nuclear matter. Such a systematic change in nucleon size would show up clearly in the gluons'

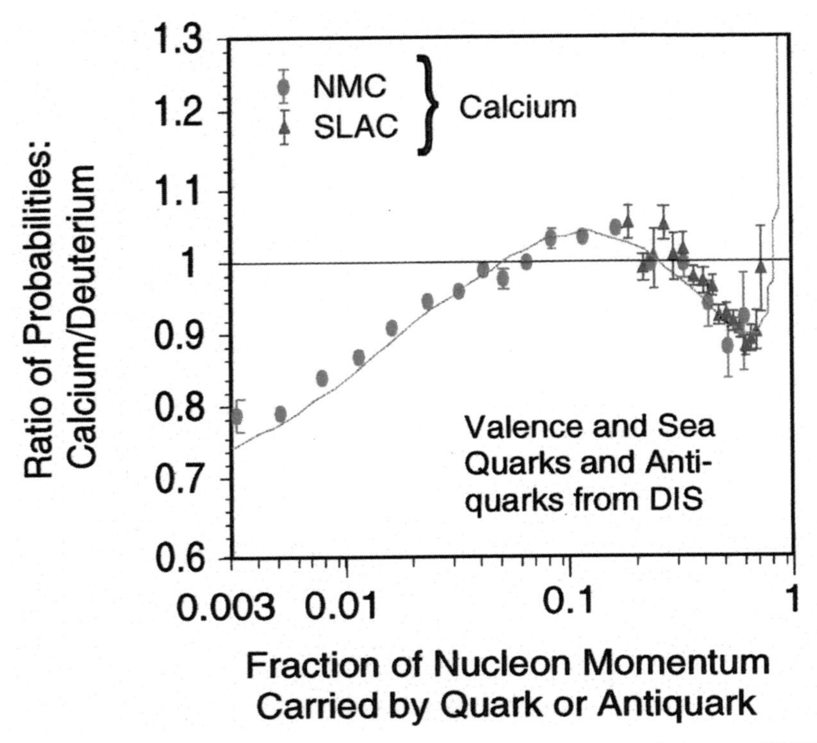

FIGURE 2.8 Results of deep-inelastic-scattering experiments carried out at CERN and SLAC that probe the distribution of quarks and antiquarks in nuclei. The results are plotted as the ratio of the probabilities for finding quarks and antiquarks with a given fraction of the average nucleon's momentum in a medium-mass nucleus versus deuterium. The sizable deviations of this ratio from unity demonstrate that the quark environment inside nuclear matter is different from that inside a free nucleon. Important nuclear effects are observed at both small momentum fractions, where induced sea quarks are expected to dominate, and at larger values, in the realm populated primarily by valence quarks. Such measurements provide one window on possible changes in hadron substructure inside nuclear matter. For example, in the green model calculation that reproduces the data, it is assumed that bound neutrons and protons have systematically different sizes than their free counterparts.

share of the nuclear momentum, which will first be measured in proton-nucleus collisions at RHIC.

Even more dramatic changes in structure, offering unique opportunities to test QCD, can occur for very fast hadrons passing through nuclear matter. During their short traversal time, quantum mechanics allows these particles to exist not just as any one of the known free hadrons, but as some combination of many of them (e.g., a combination of a proton and its various excited states, which may have a much smaller size than a free proton). Experimental searches for effects of such shrunken, transitory hadrons are of special interest, because QCD predicts that they will survive passage through nuclear matter much more readily than hadrons of ordinary size. Their survival is enhanced, according to QCD, because the various color charges are closer together inside a small hadron, and interactions with surrounding nucleons grow weaker with decreasing separation of these internal colors. This QCD prediction—that a nucleus will act as a sort of hadron "strainer"—will be tested in the coming decade by experiments that attempt to produce such shrunken hadrons in hard, high-energy collisions of electron, proton, or heavy-ion beams with nucleons in nuclei. If the prediction is confirmed, then nuclei may be used as filters to study the transition from the simplest QCD interactions to the complicated ones that dominate in the more permanent world of nucleons and nuclei. Full exploitation of such filtering possibilities is likely to require a new accelerator, capable of delivering continuous electron beams, of substantially higher energy than are available at CEBAF, or colliding beams of electrons and protons.

It is equally important to study how the interactions of lower-energy hadrons change when they are embedded in nuclear matter. The effects of the nuclear medium on nucleon-nucleon and pion-nucleon forces have been investigated intensively for many years. There is now renewed interest in the interactions of hadrons containing a strange valence quark, because they may play an important role in the high-density matter present in neutron stars, as detailed in Chapter 5. The lightest mesons that contain a strange valence quark or antiquark are called kaons. Kaons can be produced in the laboratory—and collected to form a beam—or can be implanted in nuclei by nuclear reactions that substitute a strange quark for an up or down quark. Experiments at CEBAF and other laboratories will provide information on the interaction of implanted, negatively charged kaons with the surrounding nucleons in a nucleus. It has been predicted that the partial restoration of chiral symmetry in nuclear matter, discussed above, will change this interaction at very low energy from being slightly repulsive in the free case to attractive in nuclei. If this prediction is borne out by future experiments, it could have a strong bearing on the possibility that large numbers of kaons "condense" into a single quantum state in neutron stars.

OUTLOOK

Nucleons are built from quarks and gluons. Yet nearly all of their mass—hence, the total mass of which we and our surroundings are made—appears to arise not from intrinsic quark and gluon masses, but rather from the excess baggage the quarks carry by virtue of their confinement within the nucleon. It is crucial for understanding the structure of matter to test the validity of QCD as a theory of quark confinement within hadrons. The next decade should bring enormous progress toward a meaningful confrontation of theory and experiment.

The new facilities at CEBAF and RHIC, together with continuing opportunities for selected experiments at high-energy laboratories, will dramatically improve empirical knowledge of nucleon structure. Ongoing investigations of the meson spectrum should confirm or disprove the discovery of hadrons whose primary constituents are gluons, thereby testing a prediction directly tied to the confinement mechanism of QCD. Measurements of pion interaction probabilities at low energies will reveal whether QCD leads to a self-consistent treatment of the observed violations of one of the theory's basic symmetries. This experimental progress will exploit a wide variety of accelerator facilities and state-of-the-art instrumentation. Still, the quest for ultimate understanding of the excess baggage carried by quarks within nucleons is likely to require upgrades to currently available facilities. For example, continuous electron beams of higher energy or a spin-polarized, electron-proton collider may be needed to probe correlated behavior among pairs of quarks, or spin contributions from the abundant gluons, which each carry less than a few percent of the nucleon's momentum.

In parallel with experimental progress, advances in computer performance and in theoretical techniques will fuel more quantitatively credible numerical solutions of QCD on a space-time lattice. Direct comparisons of experimental results to quantitative QCD predictions for aspects of meson and nucleon structure will have to form one of the primary testing grounds for the theory, just as the validity of quantum electrodynamics has been established in good part by quantitative accounts for observed details of atomic energy levels. On the other hand, it is unlikely that numerical solutions of QCD will be viable within a decade for systems containing more than a single nucleon. Further development of more phenomenological approaches, inspired by QCD but also guided by experimental results, will be needed to establish more firmly the QCD basis for the force between nucleons, or the structure of hadrons in dense nuclear matter.

Investigations of the evolution of hadron structure with the density and temperature of surrounding nuclear matter are just beginning. Resolution of some of the relevant issues, such as QCD effects on the passage of fast hadrons through nuclear matter, may well require accelerator facilities with beam energies and characteristics beyond those now available. Other questions, such as the occurrence of pion or kaon condensation or of a transition from nuclear to quark matter at very high densities, may only be settled by combining substantial extrapolations from laboratory experiments with successful models of stellar behavior, constrained by astronomical observations. In addressing these issues, nuclear physicists are attempting to build an essential bridge, linking the fundamental theory of nature's strongest force to the makeup of the densest objects in the cosmos, and spanning the microscopic structure of the atomic nuclei that constitute nearly all of the mass we observe around us.

3

The Structure of Nuclei

INTRODUCTION

The nucleus, the core and center of the atom, is a quantal many-body system governed by the strong interaction. Just as hadrons are composed of quarks and gluons, the nucleus is composed of the most stable of these hadrons—neutrons and protons. The question of how the strong force binds these nucleons together in nuclei is fundamental to the very existence of the universe. A few minutes after the Big Bang, the mutual interactions between nucleons led to the formation of light nuclei. These, and the subsequent nuclear process synthesizing heavier nuclei during stellar evolution and in violent events like supernovae, have been crucial in shaping the world we live in.

One of the central goals of nuclear physics is to come to a basic understanding of the structure and dynamics of nuclei. In approaching this goal, nuclear physicists address a broad range of questions, from the origin of the complex nuclear force to the origin of the elements. Among the key issues still to be resolved are the following:

- How do the interactions between quarks and gluons generate the forces responsible for nuclear binding?
- What is the microscopic structure of nuclei at length scales of the size of the nucleon? Is this structure best understood by including quarks and gluons explicitly in the treatment of nuclei?
- How are the different approximate symmetries that are apparent in nuclear structure related to the underlying interaction and how can they be derived from many-body theory?

- What are the limiting conditions under which nuclei can remain bound, and what new structure features emerge near these limits?
- What is the origin of the naturally occurring elements of our world?

Quantitative answers to these questions are essential to our understanding of nuclei; they also have a potential impact far beyond nuclear structure physics. Probes of short-range structures in nuclei can illuminate the nature of quark confinement, by exposing the extent to which quarks either remain confined to their particular neutrons or protons within nuclear matter or are shared among nucleons as electrons are shared in molecules. As yet poorly understood properties of medium-mass nuclei and of very neutron-rich nuclei critically affect the collapse and explosion of supernovae. In creating the heaviest nuclei in the laboratory, nuclear physicists are extending the periodic table of the elements and revealing deviations from chemical periodicity. Among the new isotopes they have produced in approaching the limits of nuclear stability are ones whose radioactive decay will provide crucial new tests of fundamental symmetry principles.

Progress in all these areas relies on technical advances in theoretical and computational approaches, as well as in accelerator and detector design. For example, investigations of short-range structures in nuclei have been spurred by novel developments in proton accelerators and, especially, by the advent of continuous high-energy electron beams. The role of quarks and gluons in such structures is most likely to be revealed in the lightest nuclei, for which experimental maps can now be compared to essentially exact theoretical calculations based on the picture of interacting nucleons. These calculations have been made possible by adapting the latest quantum Monte Carlo computing methods to the unique aspects of nuclear forces.

On the other hand, it is well known in all branches of physics that a direct approach to the dynamics of complex many-body systems, based on the elementary interactions between their constituents, is not always useful. For example, many properties of heavier nuclei can be accurately described using simpler approximations that retain some, but not all, essential microscopic ingredients. Deep insight into the crucial features of nuclear structure can be gained from an understanding of why such approximations work well, and of where they break down. Particular challenges are to understand the variety of collective motions of nucleons in heavy nuclei, and the fascinating phenomenon of nuclear superconductivity. Significant progress in our understanding of heavy nuclei is expected to come from advances in experimental capabilities.

Another major advance is provided by facilities producing beams of short-lived nuclei. Current understanding of both nuclear structure and nucleosynthesis is largely based on what is known of the properties of stable and long-lived, near-stable nuclei. Between these nuclei and the drip lines, where nuclear binding comes to an end, lies an unexplored landscape containing more than 90 percent of all expected bound nuclear systems, a region where many new nuclear phenomena are anticipated. As is evident from the map of the nuclear terrain in Figure 3.1, the limits of nuclear binding are poorly known at present; often, those limits are close to the regions where the processes that form the elements in stars must proceed.

In the 1996 Long Range Plan, a new experimental facility to explore nuclei near the limits of nuclear binding was identified as the choice for the next major construction project in nuclear science. Recommendation II of the present report is the construction of such a facility. Beams of short-lived nuclei will be produced and accelerated at this facility, and their reactions with target nuclei will be used to synthesize new nuclear species in uncharted territory. By elucidating the properties of these new exotic species, and enabling their use in reactions of astrophysical interest and in tests of fundamental symmetries, this new facility will provide answers to some of the most profound nuclear structure questions identified above.

NUCLEAR FORCES AND SIMPLE NUCLEI

Measuring various properties of nuclear forces and tracing their origins to the fundamental interactions between quarks and gluons has been one of the major recent goals of nuclear physics. The long-range part of the nuclear force is known to be mediated by pions, the lightest of the mesons. However, our knowledge of the short-range parts is still incomplete. When two nucleons are separated by subfemtometer distances, their internal quark-gluon structures overlap. In such cases, description in terms of the quark-gluon exchange becomes necessary.

The force between two nucleons has been studied extensively over the years by scattering one nucleon from another, and the data have been used to constrain parameters in models of the force. In the past decade, a few successful parameterizations of the low-energy nucleon-nucleon force have emerged; they offer descriptions that differ in their assumptions about short-range behavior. It is an important challenge to experiment and theory to find ways to better understand this aspect of the nuclear force, where the interface with QCD is the most critical. Such information is provided, for instance, in experiments measuring meson production in nucleon-nucleon collisions. In reactions at threshold energies, the two colliding nucleons must come essentially to rest, giving up all of their kinetic energy to produce the meson's mass. The rate of such reactions is sensitive to the strong, short-range parts of nuclear forces. New experiments aim to obtain additional information on pion production by using spin-polarized beams, and to search for the threshold production of heavier mesons. These experiments also probe meson-nucleon interactions at very low energies and provide crucial tests of QCD-based techniques for deriving the effective nucleon-nucleon interaction.

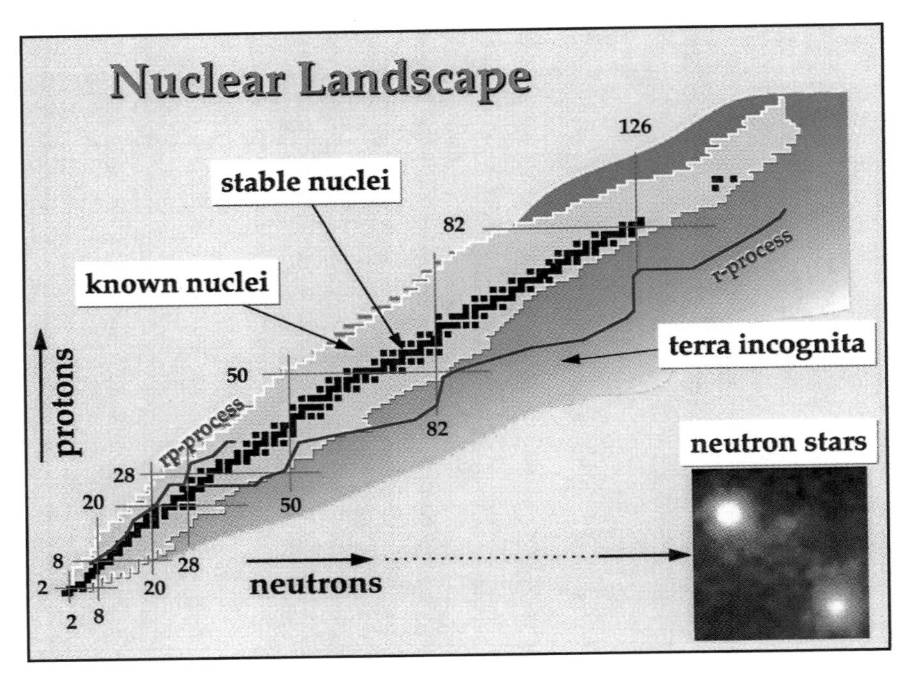

FIGURE 3.1 The bound nuclear systems are shown as a function of the proton number Z (vertical axis) and the neutron number N (horizontal axis). The black squares represent the nuclei that are stable, in the sense that they have survived long enough since their formation in stars to appear on Earth; these form the "valley of stability." The yellow color indicates man-made nuclei that have been produced in laboratories and live a shorter time. By adding either protons or neutrons, one moves away from the valley of stability, finally reaching the drip lines where nuclear binding ends because the forces between neutrons and protons are no longer strong enough to hold these particles together. Many exotic nuclei with very small or very large N/Z ratios are yet to be made and explored; they are indicated by the green color. The proton drip line is established by experiments up to $Z = 83$. In contrast, the neutron drip line is considerably further from the valley of stability and harder to approach. Except for the lightest nuclei where it has been reached, the position of the neutron drip line is estimated on the basis of nuclear models; it is uncertain due to the large extrapolations involved. Green and purple lines indicate the paths along which nuclei are believed to form in stars; only some of the dominant processes are shown. While these processes often pass near the drip lines, the nuclei decay rapidly within the star into more stable ones. One important exception to this stability plot occurs in extremely massive and compact aggregations of neutrons, neutron stars, under the combined influences of the nuclear forces and gravity.

Unique information on the strong force between hadrons can be obtained by comparing the forces between two nucleons and between a nucleon and a lambda particle in which one of the quarks is a heavier strange quark. Any difference between these forces is entirely due to the change in a single quark. The force between the lambda particle and the nucleon is being mapped with the improved experimental capabilities at CEBAF, as well as through the investigation of bound nuclear systems called hypernuclei, in which a nucleon is replaced by a lambda particle.

Even the best available parameterization of the nucleon-nucleon force cannot accurately explain nuclear binding. In order to reproduce the binding energies of the simplest light nuclei, it is essential to add three-body forces to the pairwise interactions determined from nucleon-nucleon scattering. Such three-nucleon forces are expected because the nucleons are themselves composite objects whose constituents can be distorted by an external force. A more familiar example of such a three-body force is known from the analysis of orbits of artificial satellites. In the Earth-moon satellite system, the tides induced by the moon in oceans in turn alter Earth's pull on the satellite. The nuclear three-body forces are believed to be rather weak, and it has not been possible yet to measure their small effects on the scattering of three nucleons. For now, the strengths of three-body forces have been adjusted to reproduce the binding energies of light nuclei. However, a satisfactory microscopic picture of the three-body force between nucleons is still lacking.

ADVANCES AND CHALLENGES IN UNDERSTANDING LIGHT NUCLEI

An important ongoing research effort is devoted to measuring various properties of light nuclei having up to eight nucleons. These are the simplest of all nuclei, and the first quantitative comparisons between experimental and theoretical maps of their global and short-range structure have been made. These nuclei are ideal for probing the microscopic aspects of nuclear structure, especially those related to quarks and gluons. The light nuclei also have important roles in astrophysics, elementary particle physics, and energy production. For example, most of the matter in the visible universe is in the form of these light nuclei. The nuclear physics of the Big Bang and of conventional stars like our Sun is primarily governed by the reactions between light nuclei. Nuclear fusion reactors would use some of these reactions as their energy source.

Free neutrons are unstable to radioactive decay. Deuterium (^{2}H) and helium-3 (^{3}He) are the best available surrogates for neutron targets, needed for comparative measurements of the internal structure of neutrons versus protons. A detailed understanding of the structure of these nuclei is necessary for interpreting the results of such experiments.

A direct way to probe the structure of nuclei is again through electron scat-

BOX 3.1 A Microscope to Measure the Distributions of Protons and Neutrons in the Nucleus

By bombarding nuclei with electrons from the new generation of electron accelerators and using pairs of spectrometers to detect both the scattered electron and a proton ejected from the nucleus, valuable new insight is gained into how protons and neutrons are distributed in the nucleus. The energies and angles of the electron and proton are measured with the spectrometers, and from this the energy E and the momentum k of the recoiling excited nucleus are deduced. The two-dimensional energy-momentum map yields precision information on the nucleus. As shown in Figure 3.1.1, at small values of E (the so-called valence knockout region, where the recoiling nucleus is in or near its ground state), sharp spikes appear. When E is fixed to be on one of these spikes and the distribution in momentum k is examined, the resulting pattern gives a clear picture of how the least-bound proton orbital is distributed in the nucleus, providing a powerful high-energy electron microscope for studying nuclear structure.

At high excitation energies (hundreds of MeV), the picture is much less clear. From work at intermediate energies, it is known that there is a significant probability for such reactions to occur. The violence of these reactions breaks the recoiling nucleus into fragments; theoretical studies lead us to expect that the short-range part of the nuclear force must play a major role in the process. An important goal of experimental and theoretical studies of such $(e,e'p)$ reactions at high energies is to explore this new territory, that is, the high-E, high-k part of the excitation map. Studies of $(e,e'2p)$ reactions, where an energetic electron knocks out two protons, offer an additional, promising tool for finding two protons close together inside the nucleus, an excellent measure of the short-range correlations.

CEBAF was constructed in part with these kinds of studies in mind. It is capable of the required precision measurements and exceeds the capabilities of previous high-intensity accelerators in this energy range by orders of magnitude. CEBAF is now starting to open a fascinating new window for studies at short distance scales.

There are additional new knobs to turn in electron-induced studies of nuclear structure. These involve using spin to filter out special features of the reaction. The new generation of electron accelerators all have polarized electron beams, where the spin of the electron is pointed in a direction that is controlled in the experiment. For one of the spectrometers shown in the photograph on the left in Figure 3.1.1, it is possible to measure the direction of the ejected proton's spin; in other cases, the nuclear target may be polarized, having its spin pointed in some specific direction. The control of these spins amounts to having selective knobs at the experimenters' fingertips; in favorable cases, it allows the construction of three-dimensional views of nuclear structure.

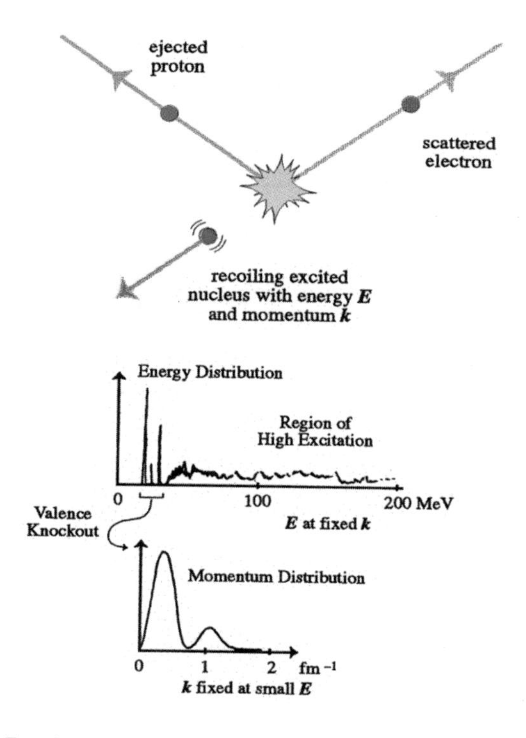

FIGURE 3.1.1 Experiments in which electrons knock out protons bound in the nucleus and both are detected in coincidence are being used to probe the distributions of nucleons in the nucleus. The photograph on the left shows an example of the equipment required for these experiments at high energies—the high-resolution spectrometers used in Hall A at CEBAF. On the right, results from previous studies with intermediate-energy electrons are shown. (Courtesy Thomas Jefferson National Accelerator Facility.)

tering. The essentially structureless electron, possessing both an electric charge and a magnetic moment, is used as a probe to map the distribution of charge and magnetism within the nucleus. As the energy of the electrons is raised, the quantum mechanical wavelength of the electron is reduced and the resolution of the maps increases.

High-energy electrons are also used to knock out the constituents of nuclei and gain valuable information. For example, by detecting the knocked-out proton or neutron along with the scattered electron, one can assess the momentum and energy distribution of nucleons in the nucleus. In particular, the data at especially large values of nucleon momenta provide crucial information on the strong forces at short distances. New insights can also be obtained by knocking out mesons from nuclei. Under certain kinematic conditions the electron can penetrate deeply within a nucleus, striking one of the virtual pions or heavier mesons being exchanged between nucleons and making it detectable.

Of particular importance to these measurements are the recently developed continuous electron beams. Previously, electron beams were largely delivered in short bursts containing many electrons. In a single burst of the beam, many particles are produced by interaction of different electrons with different target nuclei. When it is necessary to detect more than one particle produced from a single electron-nucleus interaction, the background produced in a burst often swamps signals of interest. With a continuous beam, the background can be dramatically reduced, resulting in clear identification of the desired signal (see Box 3.1).

The new experimental effort to probe light nuclei aims to fully utilize the capabilities of the recently completed continuous-beam electron accelerators and large-acceptance detectors. The ground states of most light nuclei have a nonzero spin, and many experiments propose to use spin-polarized targets to obtain three-dimensional maps of their structure. Measurements over the next 5 to 10 years will provide a detailed picture of the quantum-mechanical wave functions of two- and three-nucleon systems. Key goals are to get a quantitative measure of the small components in the wave function to pin down the role of quarks in hadrons other than nucleons and the distribution of high-momentum nucleons.

On the theoretical side, due to recent progress in computational techniques used in nuclear physics, all the bound states of these light nuclei can now be calculated, essentially exactly, from realistic nuclear forces that can interchange protons and neutrons and may flip their spins. In particular, quantum Monte Carlo methods have been adapted to study nuclei, using massively parallel computing platforms. The method is being extended to take into account leading relativistic effects in nuclear structure.

Correlations (e.g., how the separations between nucleons are distributed in nuclei) are strongly influenced by the nature of nuclear forces. Present calculations predict that neutron-proton pairs in nuclei have substantial probability to form intricate structures of toroidal (doughnut-like) and dumbbell shapes of femtometer size, as small as a single isolated nucleon. Such shapes are formed by the joint action of the short-range repulsive force and the anisotropic pion-exchange force between nucleons. A clear view of these structures can be obtained in the only nucleus with just two nucleons, the deuteron. When the deuteron is spin-polarized in one of its three possible quantum states, it has a toroidal shape; in the other two states, it resembles a dumbbell, as illustrated in Box 3.2.

The difference between the distribution of electric charge within the different deuteron spin-states has been measured only in recent years. The measurements confirm that the density in the toroidal distribution peaks at the predicted radius of about half a femtometer, making the deuteron the smallest ring-type structure known. Similar experiments being carried out at higher energies will probe the thickness of this toroidal distribution. Since the two nucleons forming the dense part of the toroidal structure are so close, their quark substructures probably overlap. Indeed, the results of another early experiment at the TJNAF indicate such an overlap.

The scattering of electrons by nuclei also provides information about the carriers of the nuclear force. This is possible because the nuclear force is mediated in part by the exchange of charged mesons, which results in a so-called exchange current contribution to the total electric current in a nucleus. In contrast, there is no comparable contribution to the electric current in an atom, because the electromagnetic force that binds electrons to nuclei is mediated by the exchange of electrically neutral photons. Important exchange current contributions have been revealed, for example, in beta decay, in the disintegration of deuterons by energetic electrons, and in the elastic scattering of high-energy electrons from ^{2}H and ^{3}He nuclei, but a comprehensive understanding of the quantitative basis of this exchange current in terms of QCD waits upon data from the new facilities.

Significant advances have been made in methods used to compute rates of reactions involving three and four nucleons from realistic models of nuclear forces. One example is the extension of exact calculations for nucleon-deuteron reactions; at higher energies, the dynamical effects of three-nucleon forces on reaction processes may become accessible to experimental delineation for the first time. A second example is that of low-energy fusion reactions in which two light nuclei fuse, with the excess energy radiated away in the form of photons. New experiments are being carried out to study these processes at very low energy with spin-polarized beams of protons and deuterons. Progress in understanding these reactions is relevant for related fusion processes in the Sun, affected by the weak, rather than the electromagnetic, interaction. These weak capture reactions contribute significantly to energy production at solar temperatures and densities, but they occur at too low a rate to be studied directly in the laboratory.

BOX 3.2 Measuring the Shape of the Deuteron

The deuteron, with just one proton and one neutron, is the simplest of nuclei beyond the proton. The nucleon density distribution in the deuteron, illustrated in Figure 3.2.1, is determined by nuclear forces; the density is large where the forces are attractive, and small where they are repulsive. It is different for the deuteron states with different spin projection because the pion exchange forces depend upon the nucleon spin orientations. The dumbbell-shaped distribution in the state with spin projection ± 1 is formed by rotating the toroidal distribution in the spin projection zero state.

These basic properties of the deuteron can be measured with a conceptually simple but technologically demanding experiment. Deuterons are bombarded with high-energy electrons, and the spin projection of the struck deuterons along the direction of their recoil is measured. Apart from small corrections, when the quantum wavelength of the recoiling deuterons is twice the diameter of the doughnut, the recoiling deuterons have only zero spin projection, and the anisotropy shown in Figure 3.2.2 has its minimum possible value of -1.4. The measured value of this wavelength then is a clear signal that the radius of the torus in the core of the deuteron is about half a femtometer. These experiments, initiated in 1984, require high-intensity, high-energy electron beams. The most recent data, taken with 4-GeV electrons, correspond to wavelengths of about one femtometer. In this region one expects to observe maximum anisotropy when the wavelength becomes equal to the thickness of the toroidal distribution.

When the nucleons are in the densest part of the deuteron, their quark structures overlap. The effect of this overlap can be seen in the cross section for disintegrating the deuteron with a high-energy photon. There are two possible processes. In the nucleon process, three strongly interacting quarks within one nucleon absorb the photon's energy. This energetic nucleon then collides against the other, causing the deuteron to break up. In the quark process, which can only occur when the quarks in the two nucleons overlap in space, the photon's energy is absorbed by all of the six strongly interacting quarks, which break apart into two energetic nucleons. The observed cross section shown in Figure 3.2.3 has been scaled such that the quantity plotted is flat for the quark process. It appears that

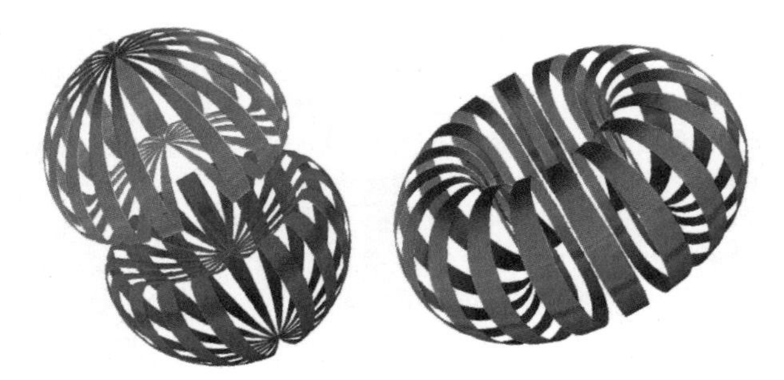

FIGURE 3.2.1 Theoretical predictions of surfaces of equal density in the deuteron in states with the projection of the spin being either one or zero along the symmetry axis. The surfaces shown are for half the maximum density of the deuteron.

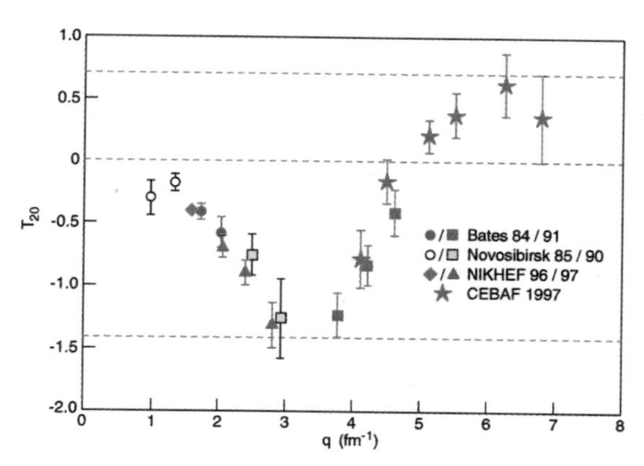

FIGURE 3.2.2 The observed difference in the way high-energy electrons scatter from the deuteron depending on the deuteron's spin is shown plotted against the recoil momentum or wavelength of the recoiling deuteron. This difference is characterized by the quantity t_{20}, which is –1.4 when all recoiling deuterons have a spin projection of zero along the direction of recoil; the other extreme is t_{20} = 0.7, when all deuterons have spin projections of one in this direction.

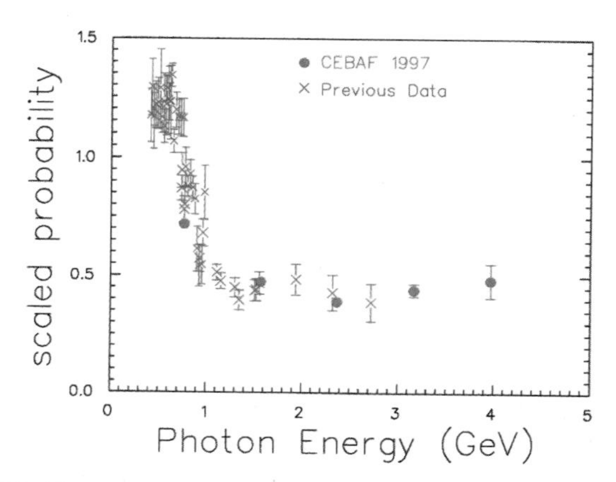

FIGURE 3.2.3 The scaled probability for the disintegration of the deuteron by a high-energy photon into a neutron and proton of equal energy is plotted against the energy of the photon. The probability is scaled such that in the quark picture it would be constant, independent of the photon energy.

the nucleon process is dominant at small photon energies, while at higher energies the quark process dominates. These experiments also need high-energy, high-intensity electron beams. The earlier data went up to 2.7-GeV photons, and in one of the first experiments carried out at the Thomas Jefferson National Accelerator Facility in 1996, these data were extended up to 4 GeV. A further extension to 6 GeV is planned.

NUCLEAR FORCES AND COMPLEX NUCLEI

The work on few-body nuclei discussed above demonstrates that there is a phenomenological theory for nuclear physics: an interaction derived from nucleon-nucleon scattering, augmented by a weaker three-body force, that is used to predict the properties of the lightest nuclei with present-day computing capabilities. Thus, even though the fundamental constituents of nuclei are quarks and gluons, there exists a simpler effective theory that accurately describes nuclei in our everyday world.

One important issue facing the field is to understand the relationship between this effective theory and QCD: Why does a model in which nucleons are interacting through a potential provide such a good approximation to the behavior of quarks and gluons in nuclei? Another question is how to extend this success in the lightest nuclei to heavier systems.

The fundamental difficulty is a numerical one: no existing computer is sufficiently fast to solve the quantum mechanics of interacting nucleons when the number of nucleons becomes large. Moreover, a complete solution of over a hundred interacting nucleons would be so complex that additional insights would be needed to interpret it and bring out its important features. Thus, physicists often must find alternatives to a direct approach to the complex many-body problem.

The Shell Model of Nuclei

The conceptual framework for heavier nuclei is the shell model, in which each nucleon is assumed to move in an average potential generated by its interactions with all of the other nucleons in that nucleus. This potential, or mean field, leads to the prediction that the quantum levels in a nucleus form shells within which several nucleons can reside. This mean field picture of nucleon motion within the nucleus explains a host of phenomena: the existence of particularly stable "magic" nuclei corresponding to completely filled shells, the low-lying excitations of such nuclei, and the collective responses of nuclei in the absorption of photons and other excitations. Maria Goeppert-Mayer and J.H.D. Jensen received the Nobel Prize for their pioneering work on introducing the concept of a shell model in the context of the nucleus.

The shell model's mean field was a significant success, and the model was soon developed further to include the residual part of the nucleon-nucleon interaction that could not be absorbed into the mean field. Theorists found that by solving the model problem of nucleons in several shells interacting through the residual interaction, a mapping (a one-to-one correspondence) emerged between the resulting shell model states and the energy levels measured in nuclei. Many properties of the actual states, especially when the nucleus is probed at appropriately long wavelengths, were found to match closely those of the corresponding shell model states.

The accuracy of this correspondence—the extent to which nucleons in a real nucleus appear to move in an average nuclear potential—is being addressed experimentally. In a series of measurements performed in the past decade, high-energy electrons were used to knock out single protons from nuclei as described in Box 3.1. For the magic nuclei in which both the protons and neutrons completely fill shells, the probability for this knockout process was found to be about two-thirds of that predicted by the simplest shell model picture, suggesting that nucleons indeed spend most of their time moving in shell-model-like orbits. The remaining one-third of the strength presumably corresponds to nucleons moving at high momentum in the nucleus and not directly included in the shell model. This occurs when nucleons pass close to one another and can feel their mutual strong interaction. One important role for CEBAF is measurement of the energy-momentum distribution of nucleons when they are not in their shell model orbitals. Such measurements will allow experimenters to see the effects of nucleon-nucleon interactions within a nucleus. The nature of these short-range interactions will also be revealed through the correlated pairs of nucleons that are occasionally knocked out of the nucleus.

Although the shell model simplifies the nuclear many-body problem, the resulting calculations are still quite daunting. When nucleons are distributed over several valence shells, the number of shell model configurations that intermingle through the residual force becomes quite large. Realistic calculations for nuclei lighter than ^{40}Ca have been feasible for some time. Even though the algorithms have been improved and the speed of computers is steadily increasing, this has produced only a modest advance, extending the calculations to $A \approx 50$.

One important recent advance in the shell model is helping to circumvent this problem. An alternative to conventional calculations has been provided by the quantum Monte Carlo techniques. While the method involves some approximations, it has been shown to reproduce accurately the results of exact, conventional shell model calculations. But unlike conventional techniques, it can be extended to heavier nuclei, where the number of configurations is enormous. One notable success of the Monte Carlo shell model is in calculating spin-flip responses in nuclei in the iron region that controls the critical rate of electron capture in the collapse of a supernova.

Another exciting direction for research involves the possibility that the shell model could become a much more quantitative many-body technique. It has been known for some years that a rigorous connection exists between the shell model and exact many-body theory. Because the connection was computationally difficult to make, the model became heavily phenomenological. But with recent advances in computing power, it now appears that those pieces of the many-body problem that are missing from the shell model—corresponding to the short-range interactions described above—can be put back in a systematic way. Results from exact calculations in light nuclei suggest that this goal might be within reach.

Mean Field Methods

Despite the exciting progress in shell model approaches, applications to heavy nuclei are still beyond our reach. These systems show a variety of collective motions, such as vibrations and rotations, involving many nucleons. A. Bohr, B. Mottelson, and J. Rainwater received the Nobel Prize for developing the nuclear unified model, which is capable of describing nuclear collectivity in terms of individually moving nucleons. But just as exact techniques in the lightest nuclei provide a bridge to the intermediate mass nuclei studied in the shell model, the shell model provides another bridge to the heavy nuclei, where other techniques are used. The effective interaction derived in shell model studies can be employed in mean-field studies of heavy nuclei. Studies based on the mean-field approach, preserving the self-consistency between the shape of the average potential and the geometry of the nucleonic orbits in it, have enjoyed remarkable success in predicting the masses, shapes, excitations, and decays of complex nuclei. The properties of this optimal average potential, such as its range and its shape (which often is not spherical), strongly depend on the number of nucleons. The effective interactions in complex nuclei are responsible for a variety of collective phenomena involving many nucleons. For instance, pairing between nucleons is responsible for the phenomenon of nuclear superconductivity, analogous to superconductivity of materials. Pairing has significant influence on nuclear dynamics, and innovative techniques have been developed to include its effects. Experimental tests of the models for complex nuclei come from low-energy facilities employing a variety of techniques.

Only recently, it has become possible to incorporate relativistic effects into the mean field description. A critical ingredient of the average potential needed to account for the observed magic numbers has been the so-called spin-orbit interaction, a force dependent on the direction of the nucleon's intrinsic spin relative to the sense of its orbital motion. The physical origin of the large spin-orbit force that is needed has been a mystery, given what is known about forces between two nucleons in free space. However, in the relativistic treatment of this problem, the magnitude of the spin-orbit interaction's strength is explained naturally.

In addition to the systematic approaches discussed above, a variety of phenomenological models of complex nuclei exist that give insight into many aspects of nuclear dynamics, such as collective rotations and vibrations. The philosophy is one familiar in other subfields, such as condensed-matter physics, where once the appropriate collective degrees of freedom are recognized, great simplicity often emerges from complex systems. Insight of this sort can be gained only by careful experimentation. Much of the rest of this chapter deals with efforts to gain such insight by examining nuclei at the extremes of mass, angular momentum, and ratio of neutrons to protons.

Limits of Nuclear Stability

One of the themes of today's nuclear science is the journey to the limits in several directions. For nuclear charge and mass this journey involves nuclei heavier than any that occur in nature or that have been produced in the laboratory; for the neutron-to-proton ratio, it involves the drip lines, the limit of binding; and for angular momentum, the journey involves the extremes of rapid rotation. Exploring the limits is expected to bring qualitatively new information about the fundamental properties of the nucleonic many-body system, about astrophysical processes and the origin of elements, and about fundamental symmetries. New, unexpected phenomena may be discovered.

The Quest for Superheavy Elements

Nuclear science started a century ago when Becquerel and the Curies discovered that not all the nuclei of atoms were stable (in other words, the phenomenon of radioactivity). This laid the foundation of nuclear chemistry and physics, and changed the face of science, technology, and medicine. The modern periodic table (as of 1998) ends with the 112th element, which was recently discovered. But the search for new elements is still an active field, and we may be due for another expansion at the upper end of the periodic table, shown in Figure 3.2.

The discovery of the properties of a brand-new set of chemical elements can answer questions of fundamental importance for science. What are the maximum charge and mass that a nucleus may attain? What are the proton and neutron magic numbers of the new elements? Where are the closed electron shells, and what are the chemical properties of such atoms of new elements?

The elements occurring naturally in the composition of Earth extend up to uranium, whose nucleus has 92 protons, but heavier elements have been synthesized in the laboratory. Like uranium, they are unstable; their absence in nature is due to their short life, as compared to the time since the heavy elements were formed in an explosive stellar event. The lifetime of uranium is long on this scale, comparable to Earth's age. Nuclei with atomic number $Z > 104$ would fission instantaneously if they were charged liquid drops governed by classical mechanics. However, the nucleus is a quantal system, and the stabilizing effects of shells allow nuclei of heavier elements to exist for a sufficiently long time that we can study their properties. According to nuclear models, there exists an entirely new group of superheavy elements, with charges around $Z = 114$ to 126, that are strongly stabilized by the shell effects. The first theoretical predictions of the island of superheavy elements came in the mid-1960s, and great effort is being made at several leading facilities to reach it. The heaviest element made and identified so far has $Z = 112$. The synthesis of that element required weeks of effort to produce a single atom.

All the heaviest elements found recently decay predominantly via alpha-

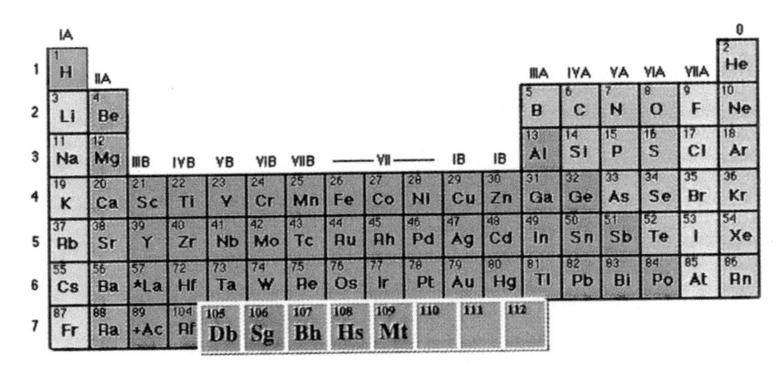
IA
0
IIA
IIIA IVA VA VIA VIIA
IIIB IVB VB VIB VIIB —VII— IB IB
1 H
2 He
3 Li 4 Be
5 B 6 C 7 N 8 O 9 F 10 Ne
11 Na 12 Mg
13 Al 14 Si 15 P 16 S 17 Cl 18 Ar
19 K 20 Ca 21 Sc 22 Ti 23 V 24 Cr 25 Mn 26 Fe 27 Co 28 Ni 29 Cu 30 Zn 31 Ga 32 Ge 33 As 34 Se 35 Br 36 Kr
37 Rb 38 Sr 39 Y 40 Zr 41 Nb 42 Mo 43 Tc 44 Ru 45 Rh 46 Pd 47 Ag 48 Cd 49 In 50 Sn 51 Sb 52 Te 53 I 54 Xe
55 Cs 56 Ba 57 *La 72 Hf 73 Ta 74 W 75 Re 76 Os 77 Ir 78 Pt 79 Au 80 Hg 81 Tl 82 Pb 83 Bi 84 Po 85 At 86 Rn
87 Fr 88 Ra 89 +Ac 104 Rf 105 Db 106 Sg 107 Bh 108 Hs 109 Mt 110 111 112
58 Ce 59 Pr 60 Nd 61 Pm 62 Sm 63 Eu 64 Gd 65 Tb 66 Dy 67 Ho 68 Er 69 Tm 70 Yb 71 Lu
90 Th 91 Pa 92 U 93 Np 94 Pu 95 Am 96 Cm 97 Bk 98 Cf 99 Es 100 Fm 101 Md 102 No 103 Lr

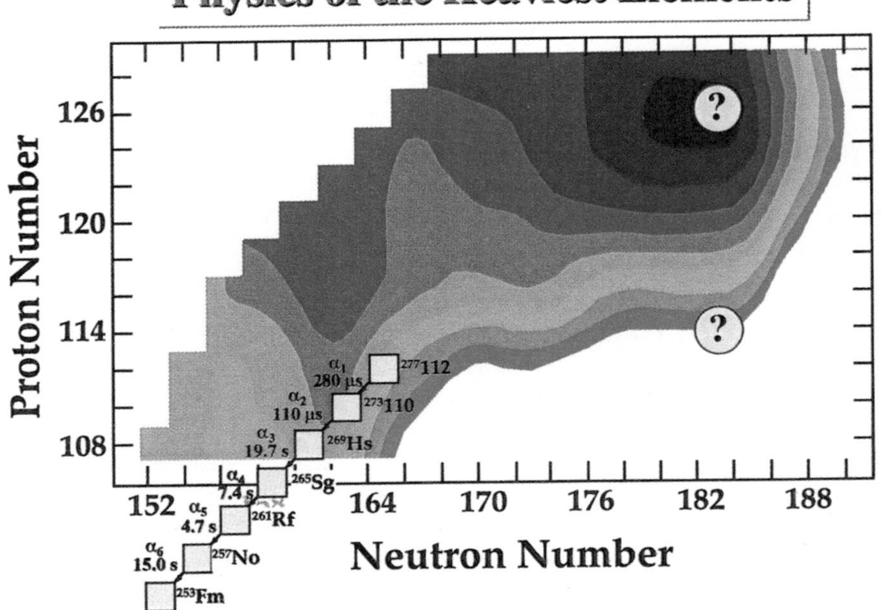
Physics of the Heaviest Elements
Proton Number
126
120
114
108
Neutron Number
152
164
170
176
182
188
?
?
α1
280 μs
277112
α2
110 μs
273110
α3
19.7 s
269Hs
α4
7.4 s
265Sg
α5
4.7 s
261Rf
α6
15.0 s
257No
253Fm

radioactivity—the emission of helium-4 nuclei—just as do the naturally occurring uranium isotopes and the lighter transuranic elements. The chain of alpha particle decays that identifies the heaviest known element as $A = 277$ and $Z = 112$ is shown in Figure 3.2. The precisely measured energies of the emitted alpha particles allow the reconstruction of the mass of the new heavy parent nucleus.

The experimental quest to synthesize nuclei with $Z > 112$ and $N \sim 184$ will not be easy, given the observed trend of decreasing yields from fusing two lighter nuclei into a heavy one. However, the superheavy nuclei are expected to live longer than the most recently synthesized elements. The long-term optimism is based on the prospects of building or upgrading facilities to produce intense beams, including beams of radioactive neutron-rich nuclei which, in combination with neutron-rich targets, may provide a path leading to the predicted superheavy elements. Experiments will almost certainly require significant advances in isotope separation and detection technology. But if they succeed in producing shell-stabilized, spherical superheavy nuclei, they will represent a major scientific triumph.

The discovery of superheavy elements would also provide crucial information on relativistic effects in atomic physics and quantum chemistry. According to calculations that include a correct treatment of special relativity, the velocity of the inner-shell electrons is predicted to approach the velocity of light as the atomic number of a nucleus approaches 173. The resulting relativistic effects

FIGURE 3.2 Top: Mendeleev's periodic table of the elements as of 1998. It includes the heaviest elements synthesized by mankind: rutherfordium, dubnium, seaborgium, bohrium, hassium, meitnerium, and the recently synthesized elements 110, 111, and 112, which are yet to be named. The chemical properties of the heaviest elements have been investigated up to nuclei as heavy as seaborgium. The strong relativistic effects cause deviations from the periodicity of chemical properties that characterizes the periodic table for lighter elements. Indeed, examples of such chemical deviations have already been observed for the elements rutherfordium and dubnium, whose properties differ from trends observed for the lighter members of their chemical families.

Bottom: Contour map of the calculated energy gain from the formation of shells in the structure of the nucleus. According to theoretical models, the region of the transactinide nuclei is connected with the region of superheavy shell stability through the valley of deformed nuclei around $N = 164$ and $Z = 110$. The experimental chain of alpha-particle decays that identifies the heaviest known element as $Z = 112$ and $N = 165$ goes through this valley. Where is the center of superheavy shell stability expected to fall? For the neutrons, most calculations predict an increased stability at $N = 184$. However, because of differing treatments of the strong Coulomb force and the spin-orbit interaction, theorists are not unanimous with regard to the position of the magic proton number. While earlier calculations yielded the value $Z = 114$, modern models, such as the one presented in this figure, favor $Z = 126$ or $Z = 120$. Both these scenarios are indicated by question marks.

cause deviations from the periodicity of chemical properties that characterizes the periodic table for lighter elements.

Toward the Limits in Neutron-to-Proton Ratio

Fewer than 300 stable isotopes, shown by black squares in Figure 3.1, occur naturally. These are the combinations of neutrons and protons that are the most stable and do not decay by the emission of radiation. In the remaining nuclei, the ratio of neutrons to protons is less optimal than in neighboring combinations, and so these nuclei decay to their more stable neighbors through weak interactions. These interactions change neutrons into protons or protons into neutrons to reach the most stable value of *N/Z* for a given total number of nucleons. The uncharted regions of the *N-Z* plane can answer many questions of fundamental importance for science: How many neutrons can be bound to a nucleus? What are the unique properties of the very short-lived nuclei having extreme values of *N/Z*? What are the properties of the effective nucleon-nucleon interaction in an environment different from that encountered in stable nuclei?

The developments in technology have been keeping up with the increased difficulties in synthesizing nuclei farther and farther away from stability. In the past several years, new nuclei having unusual *N/Z* ratios have been produced and identified. But a great deal of development remains to meet the major goal of extending nuclear structure studies out to the drip lines, where no additional protons or neutrons will be bound. The proton drip line lies relatively close to the valley of stability, hence it is easy to reach experimentally. However, because neutrons do not electrically repel one another, many of them can be added to stable nuclei before no more will be bound. The neutron drip line is therefore far from the valley of stability and its location is very uncertain, as indicated in Figure 3.1.

Unlike the processes in supernova explosions of stars, there are no direct flights to the nucleon drip lines in the laboratory by utilizing the stable or near-stable nuclei that have traditionally been available as beams and targets. Rather, one must proceed in two stages: first, conventional nuclear reactions produce radioactive species that are collected to form a secondary beam, and then that is used for the subsequent jump into the unknown region. Substantial advances in accelerator, ion source, and mass separator technology over the past twenty years have contributed greatly to the technology required to produce beams of short-lived nuclei. At present, there are only a few laboratories with only a very limited set of such capabilities. However, the success of the current programs and the interest in the new physics that they make accessible has led to developments and proposals for a new generation of such facilities worldwide.

The interest in the experimental exploration of the neutron-drip-line region is driven not only by the substantial uncertainties in its location, but also by the expectation that qualitatively new features of nuclear structure will be discovered in this exotic territory. Such features are expected for nuclei that contain a sizable number of very weakly bound neutrons. This condition should lead, for example, to a diffuse nuclear surface, in which superconducting correlations play an enhanced role.

Just before the neutron drip line, neutrons occupy orbits outside the nuclear core that are spatially extended and, from Heisenberg's uncertainty principle, have low momentum. These states, called halos, have radii that are up to several times that of the core, as is illustrated in Figure 3.3. In the limit of extremely small binding energy for a pair of neutrons to a core nucleus, giant halos will be encountered, with radii an order of magnitude larger than that of any stable nucleus.

In the surface region of heavier, very neutron-rich nuclei, one may find nearly pure neutron matter at densities much less than the normal nuclear density. The weak binding and strong pairing between surface neutrons could substantially wash out the nuclear shell structure as one approaches the neutron drip line. The concept of magic nuclei may even disappear completely in the neutron-rich extremes of the nuclear world; a possible first indication of this in the abundance of the isotopes of elements is shown in Figure 3.4. The weakening of shell structure would have a significant effect on the location of the drip line, and on the way the elements form in the neutron-rich environment of supernovae.

Improvements in the availability and intensity of radioactive nuclear beams will also greatly facilitate the study of proton-rich nuclei. Of particular interest are nuclei in the neighborhood of the $N = Z$ line (i.e., systems with nearly the same proton and neutron numbers). It is precisely in such nuclei that the fastest and best understood nuclear beta-decay processes occur. The rates of these decay processes provide stringent tests of fundamental symmetries.

The well-known phenomenon of superconductivity in solids arises from the interaction between electrons to produce correlated (Cooper) pairs. The attraction responsible for this pairing results from the interaction of the electrons with the lattice of ions in the solid. The phenomenon of superfluidity (involving superconductivity) of nuclei originates similarly from interactions between pairs of nucleons. To date, the superconducting phases associated with Cooper pairs of like nucleons—two neutrons and two protons—have been found. Heavier $N \approx Z$ nuclei may also reveal a new form of nuclear correlation from proton-neutron Cooper pairs having a unit spin like that of the deuteron.

Proton-rich nuclei offer the unique opportunity to study the nuclear system beyond the drip line. Although the protons in this region are not bound, their escape is delayed by an almost counterintuitive consequence of the repulsive Coulomb force. In these nuclei, the proton can leave the nucleus thanks to a process called quantum tunneling. In favorable cases, the tunneling of a correlated pair of protons is expected. Pioneering studies of proton emitters are already being carried out at several facilities, using stable beams and advanced detection systems.

^{11}Li: Borromean Halo Nucleus

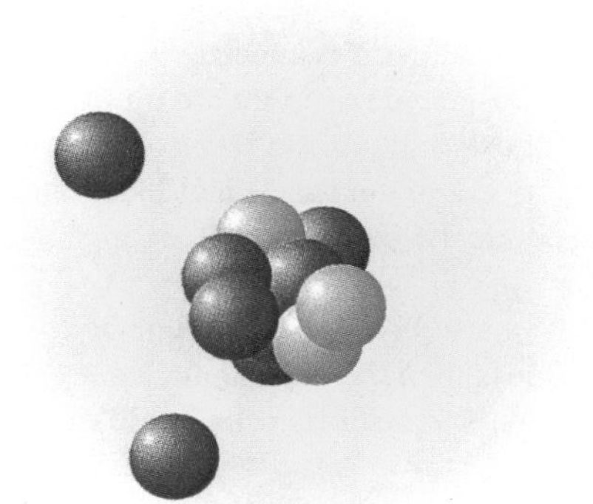

The Borromean Rings

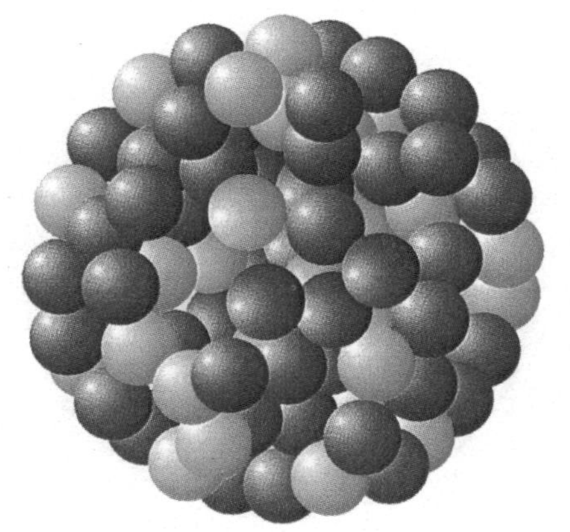

^{208}Pb: Well Bound Heavy Nucleus

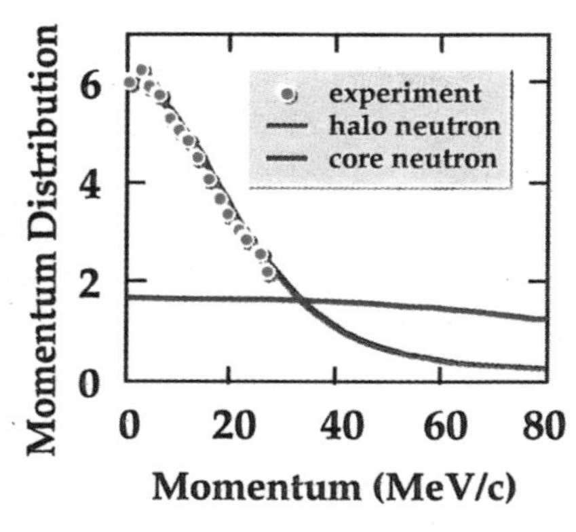

^{11}Be: Momentum Distribution

While most of our knowledge of nuclei is limited to the neighborhood of the valley of stability, early experiments on far unstable nuclei have already revealed surprises. New phenomena such as halo nuclei—with regions of nearly pure neutron matter—and growing evidence of the fragility of shell structure far from the valley of stability are just two examples. It is becoming increasingly clear that some of the cherished concepts of nuclear structure may apply only to the set of relatively stable nuclei. The advent of radioactive ion beams offers a tool to attack the basic questions pertaining to the behavior of nucleonic matter—its binding, its dynamics, and its phases.

Limits of Angular Momentum

In spite of the fact that the number of nucleons in the nucleus is rather small, and nucleonic velocities are fast, protons and neutrons can organize themselves into states exhibiting collective motion, such as rotations or vibrations. Nuclear collective rotation is uniquely interesting due to the presence of both shell structure and superconductivity.

Many nuclei emit photons characteristic of the sequential de-excitation of excited states within rotational bands. The energy levels within such a band appear to share a common intrinsic distribution of nucleons, but differ from one another in having different degrees of collective rotation. In quantum mechanics, such collective rotation can arise for systems that have a deformed shape; the characteristics of the photons emitted by the rotating nucleus provide insight into the shape and structure of nuclei, as illustrated in Figure 3.5. Some of the exciting recent discoveries in nuclear structure are associated with such rotational

FIGURE 3.3 Loosely bound halo nuclei such as lithium-11 and beryllium-11 are unique few-body systems. A paradigm of the unexpected phenomena and unusual topologies that may occur in the vicinity of the neutron drip line is the nucleus ^{11}Li (3 protons and 8 neutrons), understood as a three-body halo consisting of two neutrons and a well-bound ^{9}Li core. Interestingly, while all three constituents of ^{11}Li form a bound system when placed together, the nuclear potential is not strong enough to bind any two of them separately; hence, the name borromean nuclei. (This name comes from the Borromeo family of Renaissance Italy, whose family coat of arms pictured three rings interlocking in such a way that removing any one ring would cause all three to fall apart). Because of very weak binding, the last neutrons in lithium-11 occupy the same volume of space as the 208 nucleons in lead. Bound neutrons are not permitted to go far away from the nuclear core, by classical laws of motion; the halo structure occurs by grace of quantum mechanics. The existence of halo structure in beryllium-11 has been deduced experimentally at the Michigan State University National Superconducting Cyclotron Laboratory by studying the momentum distribution of the most weakly bound neutron (shown in the lower right of the figure). Compared to that for the well-bound neutron belonging to the beryllium-10 core, this distribution is exceptionally narrow, as expected from the Heisenberg uncertainty principle for a particle occupying an extended volume in space.

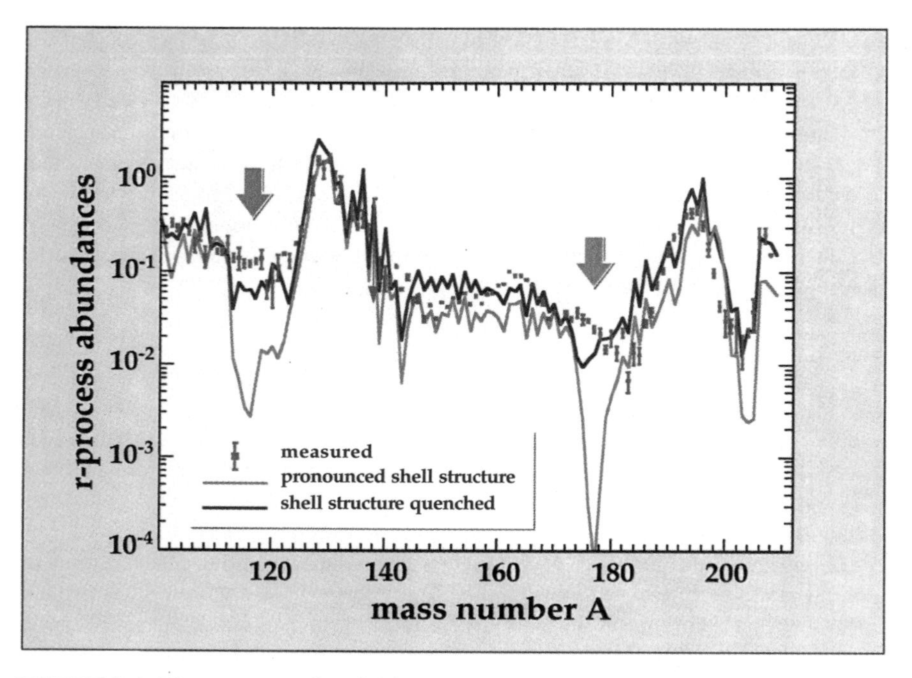

FIGURE 3.4 The structure of nuclei is expected to change significantly as the limit of nuclear stability is approached in neutron excess. Both the systematic variation in the shell model potential and the increased role of superconducting correlations give rise, theoretically, to the quenched neutron shell structure, characterized by a more uniform distribution of levels with dramatically reduced shell gaps. An important aspect of this shell quenching is its influence on astrophysical processes and on stellar nucleosynthesis. Very neutron-rich nuclei cannot be reached experimentally under present laboratory conditions. On the other hand, these systems are the building blocks of the astrophysical process of rapid neutron capture (the *r*-process, see Chapter 5); their separation energies, decay rates, and neutron capture cross sections are the basic quantities determining the elemental and isotopic abundances in nature. Consequently, one can actually hope to learn about properties of very neutron-rich systems by inverting the problem and studying the *r*-process component of the solar system abundances of heavy elements. The black squares with error bars indicate the observed *r*-process abundances for nuclei with mass numbers greater than $A = 100$. The theoretical abundances, marked by red and blue, were obtained in the recent *r*-process network calculations. They are based on models that assume that the spherical shell effects towards the neutron drip line are either similar to those in stable nuclei (red curve) or significantly reduced (blue curve). It is seen that a quenching of magic gaps at $N = 82$ and $N = 126$ can lead to a dramatic increase in the calculated abundances of nuclei around $A = 118$ and 176, in better agreement with the data. (Adapted from Bernd Pfeiffer et al., Zeitschrift für Physik A357, 253, 1997.)

bands indicating very elongated, superdeformed shapes having the longer axis twice as long as the shorter (see Box 3.3). This simple 2:1 ratio gives rise to a set of magic numbers that are different from those in spherical nuclei, and their existence is necessary for the stability of the observed superdeformed states. There is an ongoing search for even more deformed, hyperdeformed states with a 3:1 axis ratio, which are expected at the largest spins (just before the nucleus breaks up).

In a state of a very high angular momentum, the nucleus behaves like a solid in a strong magnetic field. Hence, many magnetic phenomena may occur, similar to effects already known from condensed-matter physics. One of the main challenges for the theory of rotating nuclei is to understand the nature of the dramatic spin polarization induced by fast rotation.

The spacing of energy levels in a nucleus changes from the regular, widely spaced pattern near the ground state to the more dense and random or chaotic pattern higher up. Gamma-ray spectroscopy with the new generation of detector systems, such as Gammasphere, is a unique probe of quantum chaos, roughly defined as a regime where the pattern of quantum numbers that may be used to characterize low-lying states of a many-body system is gone. Here, the important issues under study are these: At what energy does chaos set in? What are the unique fingerprints of the transition from regular to chaotic motion? The signatures for the onset of chaos are observed not only in the distribution of energy levels, but also in the properties of electromagnetic transition intensities.

The increased precision of spectroscopic tools has allowed nuclear physicists to unveil some subtle, but startling, new phenomena:

- *Identical Bands.* Identical sequences of ten or more photons are observed, associated with rotational bands in different nuclei. This comes as a great surprise: it has long been believed that the gamma-ray emission spectrum for a specific nucleus represents a unique fingerprint. Explanation of these identical patterns in a wide variety of nuclei is still lacking.
- *Magnetic Rotation.* In nearly spherical nuclei, sequences of gamma rays are reminiscent of collective rotational bands, but with a quite different character: namely, each photon carries off only one unit of angular momentum and couples to the magnetic, rather than electric, properties of the nucleons. This is a new form of quantal rotor.
- *$\Delta I = 4$ Bifurcation.* Extremely small but regular fluctuations are seen in the energies of photons emitted from some superdeformed nuclei. They could be driven by quantum tunneling motion at high angular momentum.
- *Band Termination.* Large rotational velocities can induce a gradual shape transition from the deformed state, which can rotate to the nearly spherical configuration incapable of rotational motion. Such a "death of the rotational band" has been seen in a number of nuclei.
- *Nuclear Meissner Effect.* Quenching, a total disappearance of nuclear

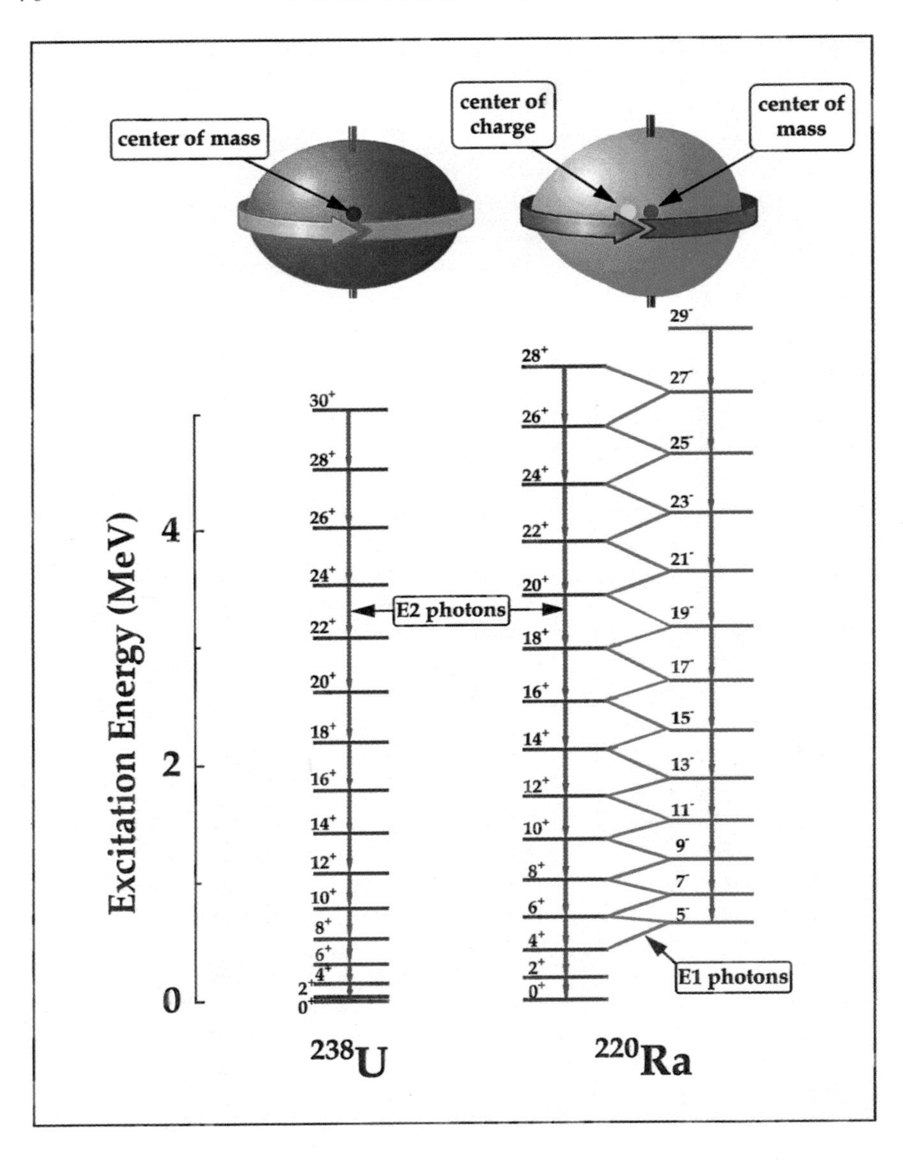
center of mass
center of charge
center of mass
Excitation Energy (MeV)
4
2
0
30+
28+
26+
24+
22+
20+
18+
16+
14+
12+
10+
8+
6+
4+
2+
0+
E2 photons
28+
26+
24+
22+
20+
18+
16+
14+
12+
10+
8+
6+
4+
2+
0+
29-
27-
25-
23-
21-
19-
17-
15-
13-
11-
9-
7-
5-
E1 photons
238U
220Ra

superconductivity, is observed at high rotational frequencies. The angular momentum behaves like an external magnetic field: it tries to align the angular momenta of nucleons along the axis of rotation, and this destroys Cooper pairs of nucleons responsible for superconductivity.

- *Symmetry Scars*. These highly excited states, such as those in a superdeformed band, represent order in chaos. The motion in these states is ordered (i.e., it is well characterized by a number of quantum numbers). The symmetry scars are embedded in the sea of many other states that can be described in terms of chaotic motion.

Advances in high-resolution, gamma-ray detector systems are also responsible for a revolution in our studies of low-spin nuclear behavior carried out at low-energy accelerators. Here, new insights have been gained into the long unresolved, but basic, questions about the mechanisms of nuclear vibrations, about nuclear superconductivity, the statistical properties of excited nuclei, and the approximate symmetries of the many-body system. For instance, long-searched-for, vibrational multiphonon states have been discovered. The very existence of these new collective modes of nuclei, both at low and high frequencies, is intimately connected with the effects of the Pauli exclusion principle. In nuclei, nature displays the enormous diversity found in the behavior of many-body systems (see Box 3.4). Exploring this diversity in experiments, discovering its new facets, and finding effective theoretical approaches to account for them remains a continuing challenge.

FIGURE 3.5 The existence of nuclei with stable deformed shapes was known early in the history of nuclear physics. The observation of large quadrupole moments led to the suggestion that some nuclei might have spheroidal shapes, which was confirmed by observing rotational band structures and measurements of their properties. For most deformed nuclei, such as ^{238}U, a description as an elongated sphere (i.e., a spheroid) is adequate to describe the band's spectroscopy. Because such a shape is symmetric, all members of the rotational band will have the same parity. However, it has since been found that some nuclei might have a shape more like that of a pear, which is asymmetric under reflection. The rotational band of ^{220}Ra consists of levels of both parities, hence the name parity doublet. Another signature of such a parity doublet is the enhanced electric-dipole radiation in ^{220}Ra due to a nonzero electric-dipole moment. Advances in high-resolution, gamma-ray detector systems are also responsible for a revolution in our study of low-spin nuclear behavior. Here, new insights have been gained on the nature of collective nuclear vibrations. In particular, vibrational multiphonon states long searched for were found experimentally. The existence of these new collective modes, both at low and high frequencies, is intimately connected with the effects of the Pauli exclusion principle.

BOX 3.3 The Champion of Fast Rotation

Rotation is a common phenomenon in nature—most objects in the universe, from the very small to the very large, rotate (Figure 3.3.1). The largest and slowest rotors are galaxy clusters. The rotation of the Andromeda Galaxy, the nearest major galaxy to our Milky Way, can be inferred from its giant spiral-shaped disk containing some hundred billion stars. Saturn is an excellent example of a deformed oblate (flattened sphere) rotator; its shape deformation is caused by a large centrifugal force. Among stellar bodies, pulsars are by far the fastest rotors; the Crab pulsar makes one revolution every 0.033 seconds! Among the dizziest mechanical man-made objects are ultracentrifuges used for isotope separation. With some modifications, the concept of rotation can be applied to small microscopic systems, such as molecules, nuclei, and even hadrons, viewed as quark-gluon systems. Atomic nuclei, with their typical dimensions of several femtometers and rotation periods ranging from 10^{-20} to 10^{-21} sec, are among the giddiest systems in nature. What makes the nuclear rotation special and interesting are quantal effects due to the nuclear shell structure and superconducting correlations.

Most rotating celestial bodies show a common behavior: at low angular momenta they acquire axial shapes depressed at the poles, like our Earth or Saturn, but at sufficiently rapid rotation, a shape transition to ellipsoidal forms having three different axes takes place. At even faster rotations, the body becomes so elongat-

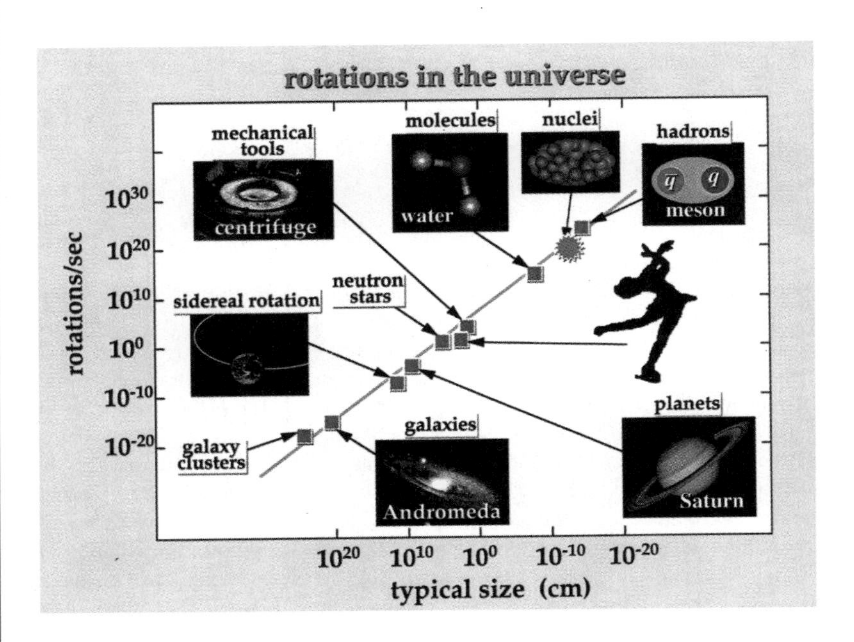

FIGURE 3.3.1

ed that it fissions into two fragments. Interestingly, the shape changes found in hot nuclei, where the shell effects and superconductivity can be ignored, seem to resemble this behavior.

The best nuclear rotators have elongated shapes resembling a football. The signature for such superdeformed states is a "picket fence" spectrum of gamma rays. The experimental spectrum of superdeformed ^{150}Gd shown in the upper part of Figure 3.3.2 was obtained in 1989 in England. The bottom, much improved, spectrum was taken at the Lawrence Berkeley National Laboratory using the new-generation, germanium-array Gammasphere with 55 detectors. The increased precision of experimental tools of gamma-ray spectroscopy has made it possible to probe new effects on the scale of a 1/100,000 of the transition energies.

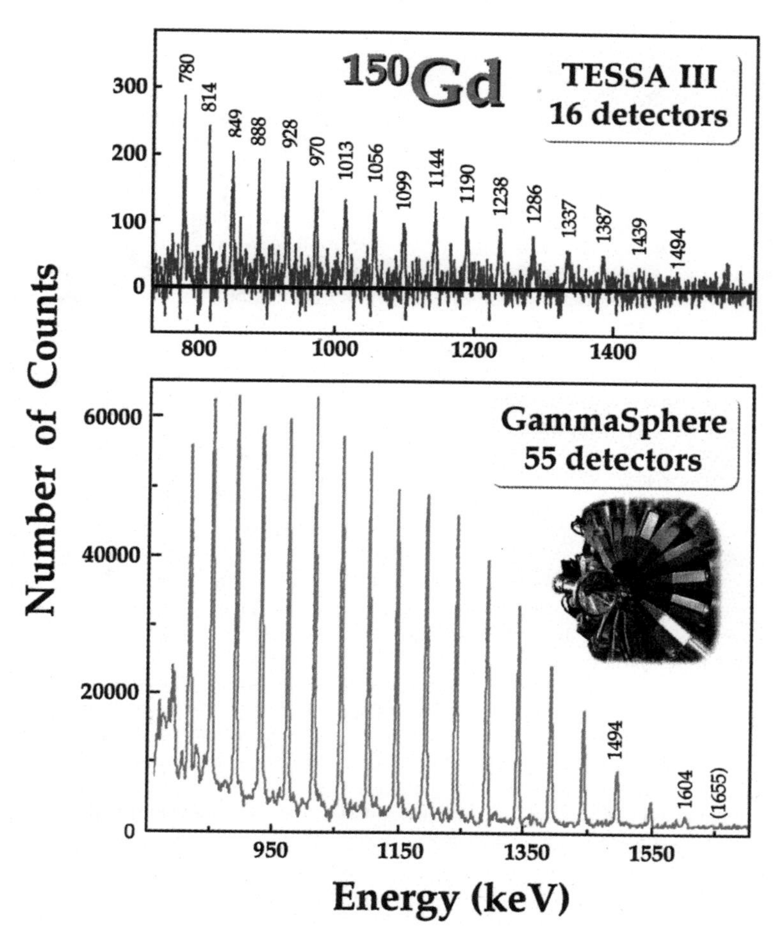

FIGURE 3.3.2

BOX 3.4 The Nucleus: A Finite Many-Body System

While the number of degrees of freedom in heavy nuclei is large, it is still very small compared to the number of electrons in a solid or atoms in a mole of gas, and as such the nucleus presents one of the most challenging many-body problems. Many fundamental concepts and tools of nuclear theory, such as the treatment of nuclear superconductivity and of nuclear collective modes, were brought to nuclear physics from other fields. Today, because of its wide arsenal of methods, nuclear theory contributes significantly to the interdisciplinary field of finite many-body systems.

From Nuclei to Molecules, Clusters, and Solids: Shells and Collective Phenomena

The existence of shells and magic numbers is a consequence of independent particle motion. The way the energy bunching of this shell structure occurs depends on the form and the shape of the average potential in which particles are moving. The electromagnetic force acting on electrons in an atom is different from the force acting between nucleons in a nucleus; this is why atoms and nuclei have different magic numbers.

Small clusters of metal atoms (typically made up of thousands of atoms or fewer) represent an intermediate form of matter between molecules and bulk systems. Such clusters have recently received much attention—not least because of their striking similarity to nuclei. When the first experimental data on the structure of such clusters was obtained, it was immediately realized that the shell-model description could be applied to valence electrons in clusters.

The nuclear shell energy and the shell energy for small sodium clusters are shown in Figure 3.4.1. In both cases, the same technique of extracting the shell correction has been used. The sharp minima in the shell energy correspond to the shell gaps. Nuclei and clusters that do not have all their shells fully occupied have nonspherical shapes. In Figure 3.4.1, the deformation effect is manifested through the reduction of the shell energy for particle numbers that lie between magic numbers. The deformation of the clusters can be deduced by studying a collective vibration that is a direct analog of the nuclear giant dipole resonance. As in nuclei, the occurrence of the deformation of the cluster implies that the dipole frequency will be split. This and many other properties of metallic clusters have been initially predicted by nuclear theorists, and their existence has been subsequently confirmed experimentally.

Interacting electrons moving on a thin surface under the influence of a strong magnetic field exhibit unusual collective behavior known as the fractional quantum Hall effect. At special densities, the electron gas condenses into a remarkable state, an incompressible liquid, while the resistance of the system becomes accurately quantized. Although the fractional quantum Hall effect seems quite different from those in nuclear many-body physics—the electron-electron interaction is long range and repulsive rather than short range and attractive—it has been shown recently that these incompressible states have a shell structure similar to that of light nuclei.

Nuclear and atomic physicists recognized early that the behavior of complex many-body systems is often governed by symmetries. Some of those symmetries reflect the invariance of the system with respect to fundamental operations, such as translations, rotations, inversions, exchange of particles, and so forth. Other

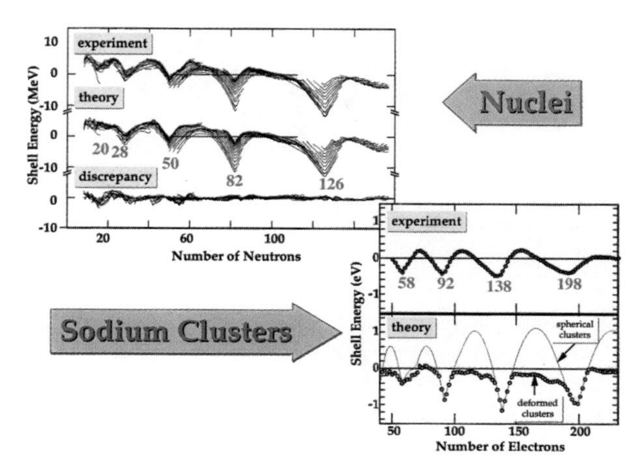

FIGURE 3.4.1 Top—Experimental and calculated nuclear shell energy as a function of the neutron number. The sharp minima at 20, 28, 50, 82, and 126 are due to the presence of nucleonic magic gaps. (From Nuclear Data Tables 59, 185, 1995 and Peter Moller.) Bottom—Experimental and calculated shell energy of sodium clusters as a function of the electron number. Here, the magic gaps correspond to electron numbers 58, 92, 138, and 198. In both cases, the shell energy has been calculated by means of the same nuclear physics technique, assuming the individual single particle motion (of nucleons or electrons) in an average potential. The reduction of the shell energy for particle numbers that lie between the magic numbers is due to deformation (the Jahn-Teller effect). (From Stefan Frauendorf and V.V. Paskevich, Annalen der Physik 5, 36, 1986.)

symmetries can be attributed to the features of the effective interaction acting in the system. These dynamical symmetries can often dramatically simplify the description of otherwise complicated systems. Symmetry-based nuclear methods have made it possible to describe the structure and dynamics of molecules in a much more accurate and detailed way than before. In particular, it has been possible to describe the rotational and vibrational motion of large molecules, such as buckyballs, hollow spheres made of 60 carbon atoms.

Chaotic Phenomena: From Compound Nuclei to Quantum Dots

With modern and powerful computers quickly processing satellite data, weather forecasts should be simple. However, the outcome of long-range weather modeling depends on seemingly trivial assumptions, and changes in the initial conditions assumed, even by a minute amount, can result in a different outcome. Even the soft flapping of butterfly wings may end in a hurricane. Predicting the final result of any chaotic system is impossible, simply because it is impossible to know the initial conditions with sufficient precision.

Corresponding to classical chaos in quantum systems, a subfield known as quantum chaos has developed. Its origins are in nuclear physics theory—specifically in the random matrix theory that was developed in the 1950s and 1960s, initially by Nobel Prize-winning Eugene Wigner, to explain the statistical properties

BOX 3.4 Continued

of the compound nucleus in the regime of neutron resonances. Today, the random matrix theory is the basic tool of the interdisciplinary field of quantum chaos, and the atomic nucleus is still a wonderful laboratory of chaotic phenomena.

Remarkable recent advances in materials science permit the fabrication of new systems with small dimensions, typically in the nanometer-to-micrometer range. Mesoscopic physics (meso originates from the Greek mesos, middle) describes an intermediate realm between the microscopic world of nuclei and atoms, and the macroscopic world of bulk matter. Quantum dots are an example of mesoscopic microstructures that have been under intensive investigation in recent years; they are small enough that quantum and finite-size effects are significant, but large enough to be amenable to statistical analysis. Quantum dots are formed in the interface of semiconductor layers where applied electrostatic potentials confine a few hundred electrons to small, isolated regions. For "closed" dots, the movement of electrons at the dot interfaces is forbidden classically but allowed quantum mechanically by a process known as tunneling. Tunneling is enhanced when the energy of an electron outside the dot matches one of the resonance energies of an electron inside the dot, leading to rapid variations in the conductance of quantum dots as a function of the energy of the electrons entering the dot (Figure 3.4.2). These fluctuations are aperiodic and have been explained by analogy to a similar chaotic phenomenon in nuclear reactions (known as Ericson fluctuations). Recently, the same nuclear random matrix theory that was originally invoked to explain the fluctuation properties of neutron resonances was used to develop a statistical theory of the conductance peaks in quantum dots. That is, a quantum dot can be viewed as a nanometer-scale compound nucleus!

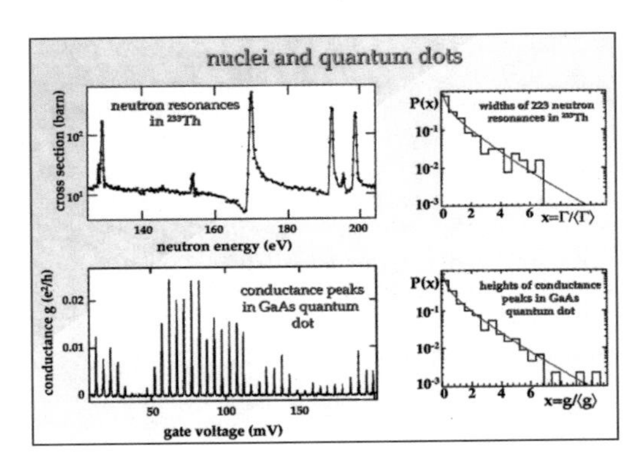

FIGURE 3.4.2 Measured neutron resonances in a compound nucleus ^{233}Th (top) and conductance peaks in GaAs quantum dots (bottom). The presence of peaks in both spectra can be explained in terms of a quantum-mechanical tunneling: whenever the energy of an incoming neutron (electron) matches one of the resonance energies inside a compound nucleus (dot), the probability for capture increases. Distributions $P(x)$ of neutron resonance widths and conductance peak heights are shown on the right. These are well described using the statistical random matrix theory, which gives the same universal probability distribution (red solid line) in both cases. (Courtesy of Yoram Alhassid, Yale University.)

NUCLEAR MATTER

While heavy and even superheavy nuclei have been discussed above, nature also provides nuclei of virtually infinite size in the dense cores of neutron stars, where hadronic matter exists as a uniform medium, rather than clumped into individual nuclei. Much of what is learned from finite nuclei—the nature of the nucleon-nucleon force, the role of many-nucleon interactions, collective excitations, and so forth—is crucial in explaining the properties of infinite matter. But many new issues, such as matter composed almost entirely of neutrons, have no counterparts in ordinary nuclei.

In the absence of Coulomb forces, the ground state of hadronic matter is a uniform liquid having equal numbers of protons and neutrons. The density of this liquid is about 2.5×10^{14} g/cc or, equivalently, 0.16 nucleons per cubic femtometer. In reality, such a liquid has a prohibitively large Coulomb energy, and thus does not exist. However, nuclei can often be regarded as relatively stable, small drops of cold nuclear matter, and some of their properties can be related to those of uniform nuclear matter.

In the cosmos, extended nuclear matter, having a large excess of neutrons over protons and containing electrons to neutralize the electric charge of the protons, occurs in the interiors of neutron stars, and briefly in massive stars collapsing and then exploding as supernovae. Studies of nuclear matter received a large impetus after the observational discovery of neutron stars in 1968, and the subsequent interest in understanding supernovae.

The dependence of pressure on density and temperature, the equation of state, is one of the most basic properties of matter. At small values of neutron excess (i.e., when the numbers of neutrons and protons are similar), cold nuclear matter is a quantum liquid. It expands on heating, and it undergoes a liquid-gas phase transition with an estimated critical temperature of approximately 18 MeV. In contrast, matter made up of neutrons alone is believed to be a gas even at zero temperature.

The binding energies and density distributions of medium and large nuclei provide information on the equation of state of cold nuclear matter at densities up to its equilibrium density and at values of the neutron-to-proton ratio up to 1.5. Currently available data are inadequate to determine the equation of state of matter with the large neutron excess that is of interest in astrophysics. Hence, extrapolations rely on theoretical predictions of the equation of state of pure neutron matter. In the next decade, a large effort will be made to study the unstable neutron-rich nuclei, using radioactive beams as discussed in the last section. This effort will provide significant additional information.

Nuclei and even large chunks of nuclear matter can undergo global vibrations in which the neutrons move against protons, or, alternatively, neutrons and protons move in unison to produce density and shape oscillations. These vibrations, conceptually similar to the fundamental vibrations of a bell or of a liquid drop, are called giant resonances. Instead of the sound waves of a bell, they emit gamma rays. These modes can be excited easily by inelastically scattering energetic particles from the nucleus, or by the shock of fusing two nuclei together. The frequencies of these vibrations are directly related to the strength of the restoring forces and to the basic properties of nuclear matter. The simplest mode, the monopole vibration, directly yields the incompressibility of nuclei and by extension of nuclear matter. Such studies have been done for some time on cold nuclei, but such new effects as the compression of the thin skin of nuclei (versus the dense inner core) are still being explored. Recently, it has become possible to excite vibrations in hot nuclei.

The short-range aspects of nuclear structure, related to nuclear forces and binding, are similar in all nuclei, and thus regarded as properties of nuclear matter. They can be studied by using high-energy electrons to knock nucleons and mesons out of nuclei from deep inside the nuclear interior. The coincidence experiments, now starting at the new generation of electron accelerators, will provide exciting new information about nuclear matter. Specifically, it should be possible to refine current understanding about the nature of nucleons bound in nuclear matter. How do they differ from free nucleons? They will also improve our understanding of the boundary between the descriptions of nuclei as bound states of nucleons or quarks. It is known from earlier, noncoincidence experiments that when the electrons are softly scattered, one obtains a relatively successful picture by assuming that they simply knock nucleons out of the nucleus. Whereas for scattering with large loss of energy and momentum, a quark-based picture appears to work. How does one go from one picture to the other? Do the descriptions coexist under some conditions? Future studies with high-energy electrons will help to elucidate these fascinating issues, especially by detecting the specific particles knocked out of the nucleus under conditions where at present only the fact that an electron was scattered is known.

OUTLOOK

The study of the structure of the atomic nucleus provides us with many insights into systems made of many strongly interacting particles. Many features of such systems are well described by amazingly simple models. The way these models emerge from the basic theory of the strong interaction is the subject of continuing study. The coming decade promises substantial progress in our understanding, by extending the study of nuclei into new domains, to the limits of their existence as bound systems. New experimental facilities and the next generation of computers are essential ingredients in this quest.

New facilities to provide exotic short-lived nuclear beams for research will open new opportunities in exploring these limits. Theoretical descriptions of nuclei far from the line of stability suggest that their structure is different from what has been seen in stable nuclei. Nuclei far from stability also play an important role in the way the universe works and how elements are synthesized in the cosmos. The properties of such nuclei are essential to a quantitative understanding of these processes. Advances in detector technology promise a wealth of new information for nuclear structure studies. New capabilities in electron scattering will provide key information on the short-distance aspects of nuclear structure.

Links in the chain that connect the fundamental theory of strong interactions of quarks and gluons to the properties of actual nuclei need to be better understood. Advances in computer technology and theoretical many-body techniques will make it possible to derive nuclear forces from the dynamics of the quarks and gluons that are confined within nucleons and mesons. New calculations will reveal the links between nuclear forces and effective nuclear interactions acting in complex nuclei. Fundamental questions concerning nuclear dynamics will be answered about the microscopic mechanism governing the large amplitude collective motion, the manifestations of short-range correlations, and the impact of the Pauli exclusion principle on nuclear collective modes. The interdisciplinary character of these studies, common frontiers with condensed matter physics and atomic physics, will be of increasing interest.

4

Matter at Extreme Densities

INTRODUCTION

The completion of the Relativistic Heavy Ion Collider (RHIC) in 1999 will open a new window on matter at the highest energy densities. RHIC will collide beams of heavy nuclei traveling at nearly the speed of light, with energies of 100 GeV per nucleon. The collisions will produce large regions of matter at unprecedented energy densities in the laboratory. One of the new phenomena expected in this regime is the deconfinement of quarks. At low energies the quarks that make up neutrons, protons, and other hadrons are always confined in groups of two or three. However, when nuclear matter is sufficiently excited by compression or heating, or both, the quarks should no longer be bound together, but should be able to move freely through the excited volume. Matter is believed to have existed in this form, a quark-gluon plasma, for the first few microseconds after the Big Bang.

The questions driving theory and experimental studies are these:

- What is the nature of matter at the highest densities?
- Under what conditions can a quark-gluon plasma be made?
- What are the basic rules governing the evolution and the transition to and from this kind of matter?

Subatomic particles, the hadrons, are composed of quarks and gluons confined within the volume of the hadron according to quantum chromodynamics (QCD). The nature of this confinement, which is a crucial aspect of the quark-gluon description of matter, is inadequately understood. At RHIC, such high-energy densities will be created that the quarks and gluons are expected to become deconfined across a volume that is large compared to that of a hadron. By determining the conditions for deconfinement, experiments at RHIC will play a crucial role in understanding the basic nature of confinement and shed light on how QCD describes the matter of the real world. These experiments are complementary to the studies of the structure of the nucleon described in Chapter 2.

An exciting theoretical challenge in studying high-energy-density matter is to understand chiral symmetry. Massless quarks possess a handedness (i.e., right-handed or left-handed); this chirality is a fundamental symmetry of QCD. In the everyday world, particles have mass. How the massless quarks turn into particles with mass is not completely understood, but the process spontaneously violates the chiral symmetry of QCD. By probing the transition between states where chiral symmetry holds and where it is broken, insight can be gained about how particles acquire their masses. Although the connection between chiral symmetry and quark deconfinement is not well understood at present, chiral symmetry is expected to hold in the quark-gluon plasma.

Information gathered from high-energy heavy-ion collisions is potentially also important in astrophysics. It will help constrain the equation of state, the equation that relates the density of matter in neutron stars and supernovae, as well as in the first microseconds of the early universe, to pressure and temperature. This information will place stronger theoretical constraints on the maximum mass of a neutron star, improving the ability to distinguish neutron stars and black holes.

A transition from normal hadronic matter to a quark-gluon plasma at high densities or temperatures is expected because, generally, as matter is heated or compressed its degrees of freedom change from composite to more fundamental. For example, by heating or compressing a gas of atoms, one eventually forms an electromagnetic plasma in which the nuclei become stripped of electrons, thereby forming an electron gas. Similarly, when nuclei are squeezed (as happens in the formation of neutron stars in supernovae, where the matter is compressed by gravitational collapse) they merge into a continuous fluid of neutrons and protons—nuclear-matter liquid. Likewise, a gas of nucleons, when squeezed or heated, should turn into a gas of uniform quark matter, composed of quarks, antiquarks, and gluons.

The regions in temperature and baryon density where the transition to a quark-gluon plasma is expected are shown in Figure 4.1. Baryons are protons, neutrons, and other particles made up of three quarks. At low temperatures and baryon densities, the system can be described in terms of hadrons, nucleons, mesons, and internally excited states of nucleons. In the high-temperature (~150 MeV or 10^{12} K), high-baryon-density (~5-10 times the density of nuclear matter) region, the appropriate description is in terms of quarks and gluons. The transition between these regions may be abrupt, as in the boiling of water—with a

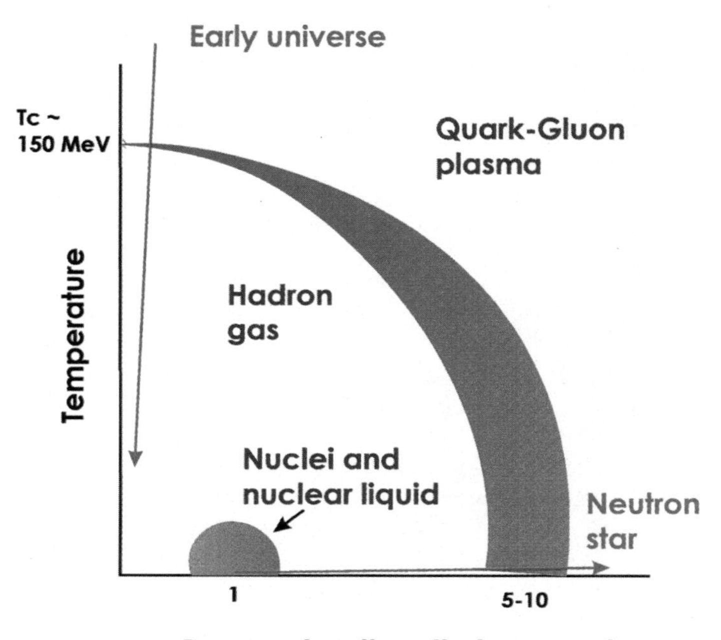

FIGURE 4.1 Phase diagram of hadronic matter. The figure shows regions of temperature and baryon density in which matter exists as a nuclear liquid, hadron gas, or quark-gluon plasma. The path followed by the early universe as it cooled from the quark-gluon plasma phase to normal nuclear matter is shown as the arrow on the left. The arrow near the bottom traces the path taken by a neutron star as it forms. Heavy-ion collisions follow a path between these two extremes, increasing both the temperature and the baryon density.

latent heat (first-order transition), without a latent heat (second order), or only a smooth but rapid crossover. The smoothness of the transition with three quarks, the realistic case, is still unclear, but the calculations show a strong change within 10 MeV of the transition temperature, 150 MeV. Under any circumstances, the physics changes strongly between the low- and high-temperature regimes.

Figure 4.1 also shows a second transition, between normal nuclei, which are liquid, to a gas of nucleons. Theory indicates that macroscopic quantities of nuclear material undergo this transition at densities below that of normal nuclear matter. The two phases should coexist at temperatures below a critical temperature of approximately 15-20 MeV. As in ordinary matter, there should be latent heat required as the system goes from the vapor, through a mixed phase of vapor and liquid, to the liquid phase. Density fluctuations and the abundances of nuclear droplets (small nuclei with nucleon number *A* in the range 6 to 50) should increase with temperature in the mixed-phase region. These phenomena have been studied experimentally, as described in Box 4.1.

From the time of the Big Bang, the early universe cooled as it expanded. For the first microseconds, the temperature was at least hundreds of MeV and matter existed as a quark-gluon plasma. Figure 4.1 shows the evolution of the early universe as a downward trajectory practically along the vertical axis of the phase diagram. The matter of the early universe had a much smaller net number of baryons than photons, about one in a billion. As the universe cooled below the critical temperature for deconfinement, the primordial plasma coalesced into hadrons. The quarks found partners and became confined into nucleons and mesons and eventually into the nuclei we observe today. If the transition is first order, droplets of hadrons formed in the middle of the plasma similar to the way water precipitates into rain drops.

Such density inhomogeneities could have enhanced the abundance of the elements Be or B. They may even have led to formation of strange quark matter nuggets or planetary-mass black holes, which could account for some of the so-far-unobserved dark matter in the universe. If the transition is relatively slow cosmologically, then the hadron formation process could contribute to the entropy, or disorder, observed as the number of photons in the universe. These features of the universe depend on how the transition from the quark-gluon plasma to hadronic matter took place.

In neutron stars the properties of matter under extreme conditions also play a crucial role. For example, our present lack of knowledge of the properties of matter at densities beyond twice that of nuclei is reflected in uncertainty about the maximum mass of neutron stars. RHIC experiments, by providing information on the equation of state, should help us determine possible states of matter in neutron stars. For example, neutron stars may, at a density as low as a few times nuclear-matter density, contain a mixed state consisting of droplets of quark matter immersed in ordinary hadronic matter fluid. If their central density rises to five to ten times that of nuclear matter, they may have quark-matter cores in their deep interiors. One cannot even definitely rule out, without further data, the possibility of a distinct family of quark stars with higher central densities than those of neutron stars.

ULTRARELATIVISTIC HEAVY-ION COLLISIONS

To date, experiments to study matter at very high energy density have been performed at several beam energies. High baryon density can be achieved by colliding nuclei at an energy where they just barely stop one another, as recent experiments have shown. The Brookhaven AGS accelerator has been an optimal tool for studying extremely dense matter. The CERN SPS, with higher energy, provides somewhat lower baryon density but higher energy density, making addi-

BOX 4.1 The Liquid-Gas Phase Transition in Nuclear Matter

Nuclear matter is expected, theoretically, to undergo a transition from a liquid to a gaseous phase at densities lower than those inside normal atomic nuclei. This phase transition should occur at a temperature of 1.7×10^{11} K or 15 MeV. Just as in an ordinary liquid like water, the nuclear-matter phase change should occur at a constant temperature if the pressure is not varied. During the transition, nuclear matter exists as a mixture of the liquid and gaseous states. Density fluctuations should occur in this mixed phase, varying the abundances of nuclear droplets—small nuclei with 6 to 50 nucleons—that are produced. Unlike the case of ordinary liquids, however, quantum mechanics plays an important role in nuclear matter and its phase transitions.

The nuclear liquid-gas phase transition can be studied by observing the disassembly of the finite nuclear systems produced by colliding atomic nuclei at laboratory accelerator facilities. Experiments measure the probability of decay of the excited system into multiple light nuclear fragments, or droplets of nuclear liquid. Experiments have been devised to measure these fragments and distinguish between the simultaneous fragment emission expected from a phase transition, and evaporation of nuclei with 5 to 60 nucleons over a longer time scale from the surface of larger nuclei. Identifying those collisions where bulk multifragmentation

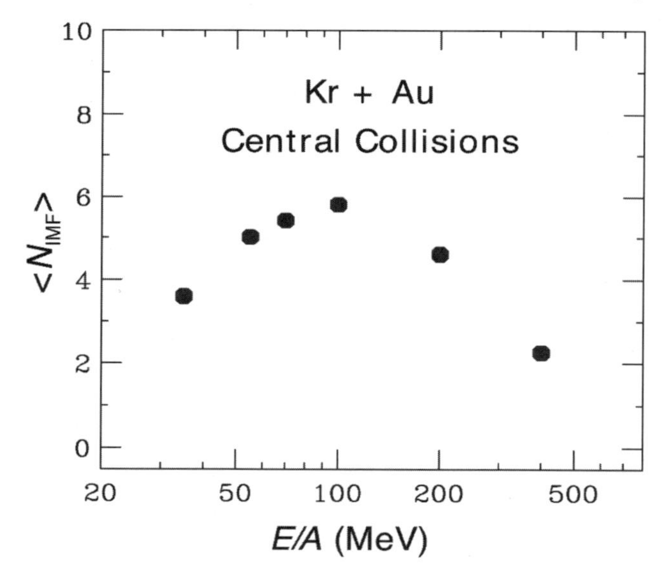

FIGURE 4.1.1 The average number of small nuclei, or intermediate-mass fragments, emitted in collisions of Kr + Au at different projectile energy per nucleon. The peak around 100 MeV/nucleon collisions suggests that a liquid-vapor phase transition takes place.

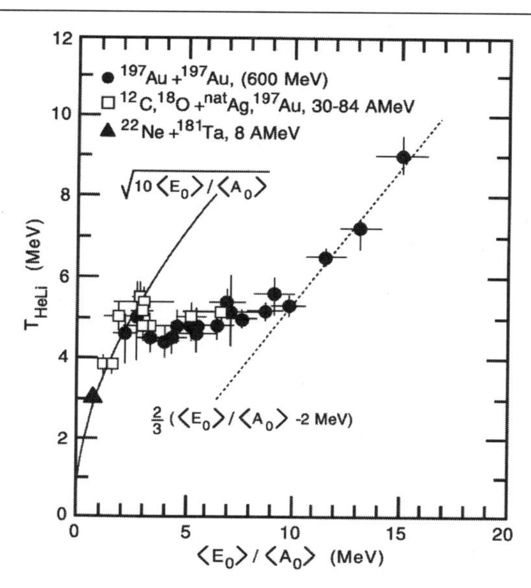

FIGURE 4.1.2 The "caloric curve" of nuclear matter. The temperature of the system created in the collision of heavy ions at intermediate energy is plotted as a function of the excitation energy available. Both quantities are inferred rather than directly measured—the temperature from the relative abundances of different nuclear isotopes, and the excitation energy from the number and energy of observed particles. Because the temperature and excitation energies are derived and rely upon theoretical assumptions, there remain large uncertainties on their values. Nevertheless, the initial rise, followed by a region of constant temperature with increasing excitation energy, ultimately followed by a second temperature rise, is the behavior of a system undergoing a first-order phase transition. More sophisticated experiments will show whether this trend actually occurs in nature or is an artifact of the way the temperature and excitation energy are currently inferred from the data.

occurs should allow extraction of the thermodynamic properties of the liquid-gas phase transition.

Some of the experimental evidence for the phase transition is shown in Figure 4.1.1, which illustrates the yield of light nuclei in collisions of Kr beams of different energies on a stationary Au target. At energies between 50 and 100 MeV per projectile nucleon, light nuclei are abundantly produced. Careful analysis of the data indicates that the excited systems decay in less than 3×10^{-22} sec, a time much less than that required for sequential emission of light nuclei, with equilibrium being reestablished between each successive step.

Other intriguing evidence that the mixed phase may be prepared in the laboratory is shown in Figure 4.1.2. For excitation energies of 2 to 10 MeV, the temperature remains nearly constant at 4.5 MeV (about 5×10^{10} K), suggesting an enhanced specific heat as in a phase transition. This figure is a encouraging step toward observing the phase transition; however, experimental uncertainties in the determination of the temperature and excitation energy scales are significant and must be improved before thermodynamic quantities can be reliably extracted.

tional observables accessible. The RHIC collisions will have ten times the energy of the collisions at the SPS. A very high energy density with a net baryon density of nearly zero is expected.

Experiments have been performed with beams of light (sulfur and silicon) and heavy (lead and gold) nuclei. These experiments have mapped how the collisions evolve in time and change character with changing energy. They have shown which observables must be measured and how such measurements are best done, given the large number of particles produced. The program has included careful comparisons to proton-proton and proton-nucleus collisions to guide the search for evidence of new physics. These experiments have revealed several surprises, as discussed below.

Collisions at increasingly higher energies create increasing numbers of mesons, as the incoming energy of the projectile is transformed into the mass carried by these mesons. Eventually the bombarding energy becomes high enough that the nucleons can no longer completely stop each other. Then the nuclei pass through one another, becoming highly excited internally, and leave the physical vacuum between them with a great deal of energy. This excited vacuum contains quarks, antiquarks and gluons (see Box 4.2). Such nuclear transparency becomes important in the highest-energy collisions. The nuclear fragmentation regions, which recede from each other at nearly the speed of light, contain essentially all the baryons of the original nuclei. The central region with very few baryons is of special interest, because it resembles the hot vacuum of the early universe. Although the hot, dense phase of the collision does not live long, the initial temperature and density reached are very high indeed.

RHIC offers a unique opportunity—the matter created in RHIC collisions will redefine the energy density and temperature frontiers. The matter is expected to exist much longer than the hot matter studied in experiments to date.

STOPPING

The degree to which colliding nuclei stop each other determines the energy available to excite the central region between the nuclei. The excitation energy heats the matter and produces particles. If a large number of particles is produced in a small volume, this high density creates pressure, causing the blob of matter to expand collectively. Experimental data on nucleon distributions and the amount of energy produced transverse to the beam direction allow one to infer the extent of stopping in collisions. Data at 14 GeV/nucleon for the heaviest nuclei are consistent with the maximum possible energy deposition, or "full stopping." At the SPS, the collisions are more than three times as energetic, and a considerable fraction, but not all, of the available energy is found to be transferred to other degrees of freedom.

Calculations that reproduce the experimentally observed stopping indicate maximum baryon densities about 10 times that of normal nuclear matter in

14 GeV/nucleon collisions, and up to 6 times at the SPS. The corresponding energy densities are about 10 times that of normal nuclei and 2 to 5 times the energy density inside a hadron. Although a smaller fraction of the incoming energy is stopped in the higher-energy collisions, the energy density is higher and the baryon density lower than at the AGS. These estimates indicate that collisions that are being studied at existing facilities may be reaching the threshold of the deconfinement transition, and signatures of the new phase may be observable. These signatures and experiments looking for them are described below. At RHIC, the stopping of the nucleons will be much less, but because the collision starts out with 10 times as much energy as current collisions at CERN, the heating and energy density achieved will, it is predicted, be at least twice that in present collisions. Consequently, experiments will no longer need to peer across the threshold but will be able to study a deconfined state thoroughly.

EVOLUTION OF COLLISIONS

Very high-energy heavy-ion collisions proceed through a number of different stages, shown schematically in Figure 4.2. In the first stage, the two colliding nuclei penetrate one another. The quarks and gluons composing the nuclei collide and transfer a large amount of energy from the projectile to the vacuum, energy later observed as new particles (Figure 4.3). This stage of the collision, lasting about 3×10^{-24} sec, is short because of the relativistic contraction of nuclei moving nearly at the speed of light.

From the very energetic or "hard" collisions among the quarks and gluons, additional gluons and light and heavy quarks are produced. These new quarks and gluons, along with those present initially, undergo a cascade of further collisions, which slows down or stops the nuclei, as discussed above. The techniques of perturbative QCD, developed to describe $e^+ - e^-$ and proton-antiproton collisions, have been applied to describe this cascade. The calculations predict very high initial temperatures, 2 to 3 times that required for deconfinement, and copious gluon production. These gluons should thermalize in less than the time required for the nuclei to interpenetrate, sharing their energy equally among themselves. This process would produce a quark-gluon plasma in local thermal equilibrium that lives long enough to generate detectable signals. The hot and dense stage of the collision, where the plasma is expected to exist, lasts only about three times as long as the interpenetration stage; experiments will have to sample this stage for evidence of the existence and properties of the quark-gluon plasma. The predicted signatures are described below, along with some intriguing results from existing lower-energy experiments.

The hot and dense phase then cools and expands, and below the deconfinement temperature and density the quarks and gluons condense into a gas of hadrons. The hadronization transition is expected to take place around 10 to 30×10^{-24} sec after the nuclei began to collide. It is at this point that most

BOX 4.2 Probing the Vacuum

Although one usually thinks of the vacuum as space with nothing in it, quantum theory tells us that the vacuum is not really completely empty. It contains a "sea" of quark-antiquark pairs. However, these particles are present in the vacuum only fleetingly, as fluctuations of the fields generated by other particles. Normally, but not always, these fluctuations are too small to be observed.

Nuclear physicists will explore the vacuum by heating it up to an extremely high temperature (some 1,500 billion degrees!) by colliding pairs of heavy ions at very high energies. The vacuum becomes polarized by the large energy density, making the sea of quark-antiquark pairs visible as discussed in the section describing the structure of hadrons. When two nuclei collide, they pass through each other, converting some of their energy to heat, heating themselves as well as the tiny piece of vacuum between the two retreating nuclei. Theory predicts that such heating will create conditions comparable to those in the first millionth of a second after the Big Bang.

The first panel in Figure 4.2.1 shows two heavy nuclei approaching each other at velocities near the speed of light. At these speeds they appear flattened because of the Lorentz contraction. Immediately following the collision, the two nuclei have formed a hot region between them, as shown in the second panel. In this hot region, there will be fluctuations of the color field that governs the interactions of quarks and gluons. A number of interesting phenomena should take place.

The color field will produce quark-antiquark pairs, converting collision energy into particles. Because of the high temperature, a very high density of quarks and antiquarks, as well as the gluons they exchange when they interact, will build up. At high densities, the quarks should no longer be confined in the particles we normally observe. Instead, they will roam freely over the hot zone, forming a quark-gluon plasma. The plasma will radiate photons and lepton pairs, such as electron-positron or muon-antimuon pairs, as indicated in the third panel. The leptons and photons are not affected by the strong force among the quarks as they escape from the quark-gluon plasma; thus, they serve as messengers carrying information about the plasma's properties.

Following the collision, the quark-gluon plasma cools and changes back to the usual hadronic phase of matter. A large number of particles, primarily hadrons (baryons and mesons), emerge from the collision. The hot vacuum cools back to its usual, seemingly empty state. However, the particles produced in the collision are left over, as shown in the bottom panel, allowing experiments to "peek" into the structure of the vacuum.

of the particles shown in Figure 4.3 come into existence. The matter is still highly excited, and the density of the hadrons is large. The hadrons scatter from one another, maintaining the pressure and causing further expansion and cooling. The multiple scatterings of the particles tend to partition the available energy equally among them and keep the system in equilibrium. Eventually, the system is sufficiently dilute that the hadrons cease colliding and travel outward without further disturbance.

The number of particles at the end reflects the energy deposited in the colli-

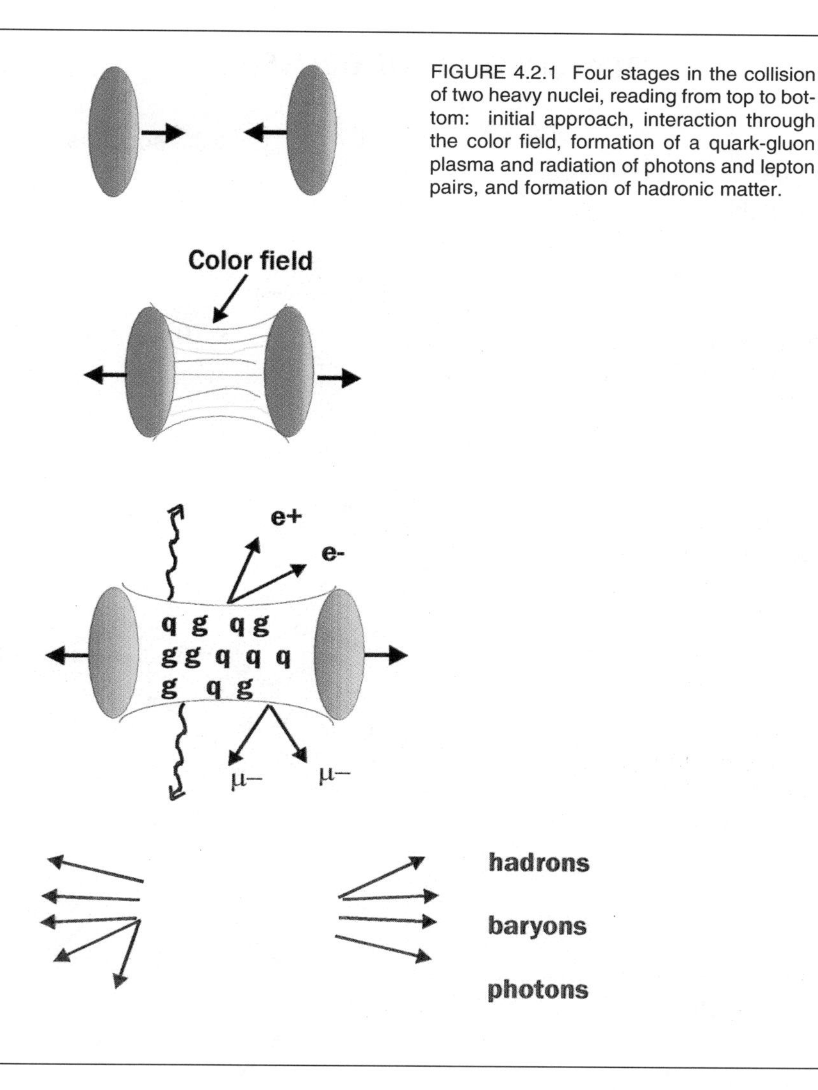

FIGURE 4.2.1 Four stages in the collision of two heavy nuclei, reading from top to bottom: initial approach, interaction through the color field, formation of a quark-gluon plasma and radiation of photons and lepton pairs, and formation of hadronic matter.

sion. This energy rises with the energy of the beam, as illustrated by Figure 4.3. The thousands of particles emerging from the space normally subtended by one nucleus indicates an extremely high energy density in that space. The detection of this large number of particles requires sophisticated and ingenious detectors with high granularity. The complexity of the detectors is comparable to that of the largest existing detectors in high-energy physics; however, the nuclear experiments must identify a larger range of particles, and they must retain sensitivity to rather low-energy particles. The requirements of measuring high-energy

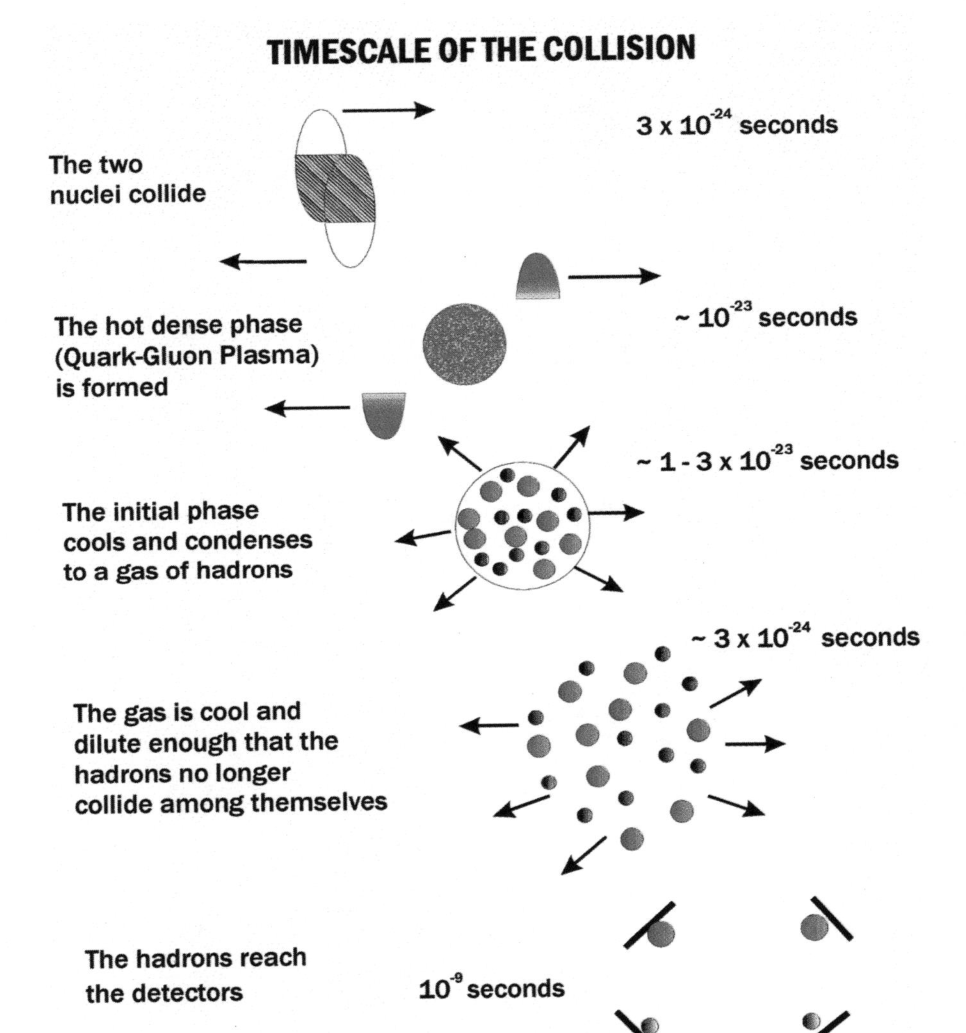

FIGURE 4.2 Schematic illustration of the different stages in a heavy-ion collision. The stages include (a) initial interpenetration of the colliding nuclei; (b) the hot dense phase where the quark-gluon plasma should exist; (c) the hot hadron gas phase; (d) freezeout of the hadrons, where they no longer scatter among themselves; and (e) the point when the hadrons reach the detectors. The timescale for each stage is indicated on the right side of the figure.

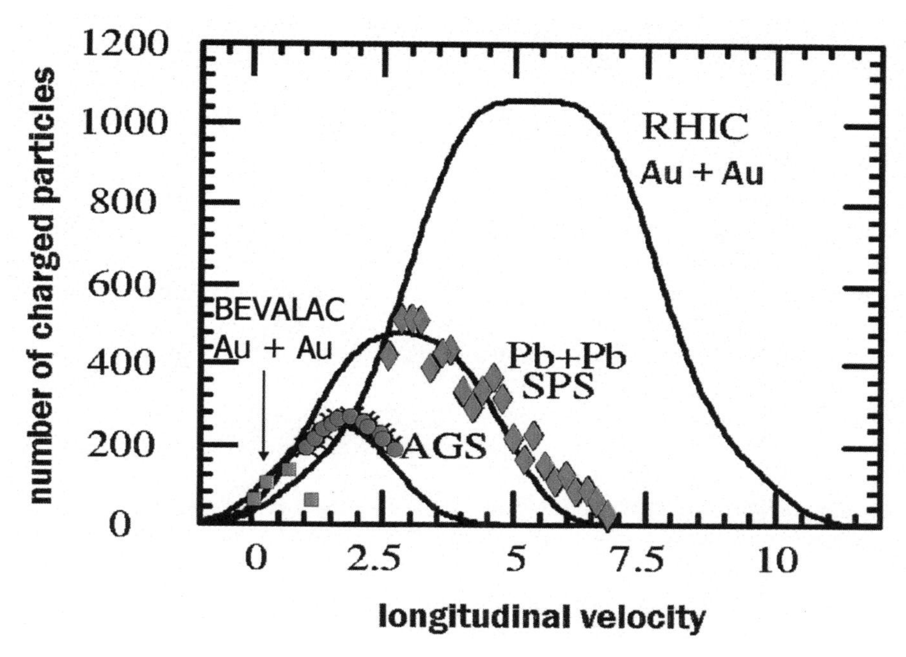

FIGURE 4.3 The number of charged particles produced in Au + Au (or Pb + Pb) collisions is plotted as a function of the longitudinal velocity of the particles. The increase in the number of particles produced from modest energy (1 GeV/nucleon beam) at the Lawrence Berkeley National Laboratory Bevalac through the AGS at Brookhaven National Laboratory and the CERN SPS is dramatic. The uppermost curve shows the number of particles predicted to be produced with the very high energy at BNL's Relativistic Heavy-Ion Collider.

collisions of very heavy ions have driven new developments in detectors, readout electronics, and data acquisition technology.

Hot Dense Initial State

The arrangement of the quarks and gluons making up the nucleons inside a nucleus has been studied with energetic electron, positron, and proton beams. New results from the HERA electron-proton collider in Germany show an abundance of gluons carrying a small fraction of the nucleon momentum. These results indicate that in addition to the quark collisions, an even larger number of collisions among gluons will occur at RHIC, causing a hot gluon gas to be formed. A hot gluon gas has never before been created, and RHIC will offer the first glimpse of such matter. The dominance of gluons causes special signatures, which can be looked for experimentally. The temperature and corresponding energy density should be significantly larger at RHIC than for AGS or SPS collisions, and the lifetime of the hot dense state should be longer.

The gluons participate in a rapid cascade of subsequent collisions. The cascade drives the central region toward thermal equilibrium on a time scale comparable to the interpenetration time of the colliding nuclei, producing exactly the conditions needed to form a quark-gluon plasma in local thermal equilibrium. Such equilibrium is unlikely in isolated nucleon-nucleon collisions because the density of gluons is not sufficient to produce the fast cascade. The hot, gluon-rich initial state should produce thermal photons, which survive the subsequent stages of the collision and enter the detectors unchanged. Heavy-ion collision experiments look for these photons. Another result of the gluon-rich initial state is the fusion of gluon pairs to produce many charmed quark-antiquark pairs, which are observable at the end of the collision via the charmed mesons they form.

In the higher energy heavy-ion collisions planned for at the LHC at CERN, the gluon densities are predicted to be even greater. Though this density should further increase the temperature and lifetime of the plasma, the LHC energy causes production of quarks and antiquarks very early in the collision as well. The initial state is thus quite complex, and the number of particles that must be detected in the final state is correspondingly higher.

HADRONIC RESCATTERING AND FREEZEOUT

The quark-gluon plasma is expected to exist in the hot, dense stage of the collision, and this stage is consequently of primary interest. However, the evolution to the finally observed hadrons must be understood in order to interpret the results of the experiments.

After the hot dense region expands sufficiently, the hadron gas becomes so dilute that scattering among the components ceases and the hadron gas "freezes out." Following this freezeout, the various particles freely fly out of the collision region, reaching the detectors nanoseconds later.

The freezeout volume can be studied by measuring the correlations of the momenta of identical particles. This technique, which relies on quantum wave mechanics, is borrowed from astronomers, who developed the use of intensity correlations to measure the size of distant stars; the spatial range of the observed correlations is related to the size of the emitting star. Similarly, in nuclear collisions, the correlations between pairs of mesons carry information about the size of the region emitting the mesons, as shown in Figure 4.4. The source sizes inferred exceed the size of the incoming nuclei. Furthermore, the dependence of the correlations on the momenta of the particles indicates that the system is expanding. Detailed measurements of both the two-particle correlation functions and the momentum spectra of single hadrons at RHIC energies should pin down the expansion velocities and the temperatures at which the hadronic phase freezes out.

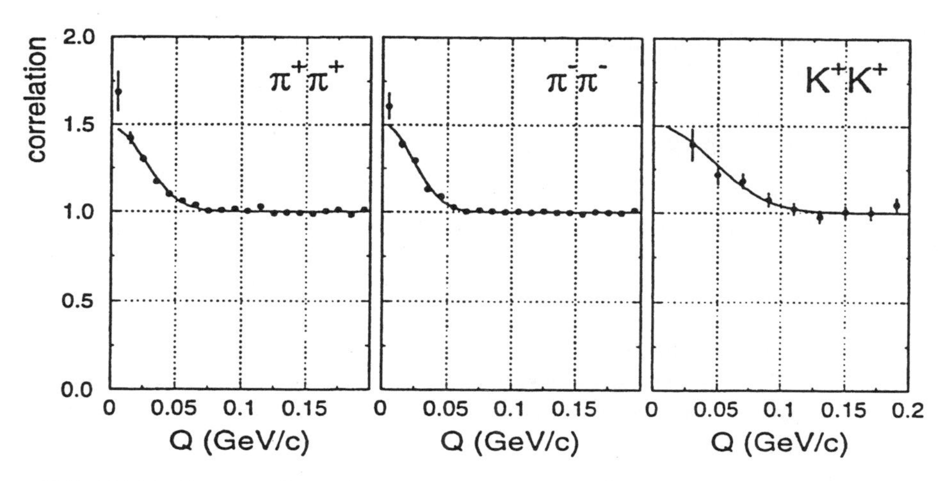

FIGURE 4.4 The rate of detecting identical meson pairs from a high-energy collision of two gold nuclei as a function of their relative momentum Q. The technique can measure a diameter of about one-trillionth of a centimeter by using the uncertainty principle to convert the width in Q of the measured peak into the size. (Courtesy of E877 collaboration.)

THERMAL DESCRIPTION OF THE FINAL STATE

The latter phase of the collision, after the quarks coalesce into hadrons, appears to behave as a system of hadrons in thermal equilibrium. The equilibrium is unlikely to span the entire central region in the collision, as the hadrons scatter primarily with nearby neighbors. Consequently, equilibration is local rather than global; however, the size of the equilibrated region is large enough to include many particles.

Experimental measurements that characterize the hadronic system include average momenta of the particles transverse to the beam. Measurement of the spectral slope, or falloff in the number of particles with increasing transverse momentum, as a function of the collision centrality—how nearly head-on the nuclei collide—permits determination of the temperature and the extent of chemical equilibrium. From such measurements, a large latent heat associated with a first-order phase transition should be observable experimentally.

In the data available at present, the production rate of different particle types is consistent with a system in local thermal and chemical equilibrium. It is possible to fit the data with this assumption and determine the point at which the particles cease to interact. The fit indicates that freezeout in collisions at 14 GeV/nucleon occurs in the temperature range 120 to 140 MeV, with baryon density 30-40 percent that of normal nuclear matter. Analysis of higher-energy data reveals freezeout at similar temperature and baryon density. One of the first experimental results awaited from RHIC is how the higher initial temperature affects the freezeout.

SIGNATURES OF QUARK-GLUON PLASMA FORMATION

Determining the nature of high-density matter and identifying the formation of a quark-gluon plasma are the major experimental challenges at RHIC. Ideally, the experiments must show that some features of the data cannot be present without a quark-gluon plasma. A number of such signatures have been predicted and are listed in Table 4.1. This section describes several of the table entries in more detail and summarizes what is known from experiments thus far.

Typically, an experiment looks for multiple signatures as a correlated change in the observables, such as those shown schematically in Figure 4.5. Varying the conditions of the collisions (energy and size of the colliding nuclei, for example) will allow experimental determination of the energy density required for the deconfinement transition and will allow its properties to be studied. Putting together information from the different signals will shed light on how confinement works and how chiral symmetry is restored. Even short-lived or partially equilibrated plasma should exhibit properties different from normal hadronic matter and, thus, be detectable.

The first five signatures in Table 4.1 are detectable by measurement of electrons and muons emitted from the collisions. These leptons are frequently emitted in pairs, one positive and one negative; the invariant mass of the pair is calculated from the measured energy of the two leptons. As leptons and photons interact electromagnetically, they exit the colliding system virtually unaffected by the surrounding hadronic matter. This allows them to provide information about the early stages of the collision. Figure 4.6 shows schematically how the dilepton mass distribution would be modified, should all of these predicted signals occur.

A striking signature of quark-gluon plasma existence is the suppression of J/psi labeled region 4 in Figure 4.6. The interaction between quarks is screened by other quarks in the deconfined state, so quark bound states such as the J/psi cannot survive in the plasma. Bound states with a large radius are dissociated first, while the small ones, like the upsilon (at a mass of about 9.5 GeV, beyond the range shown in Figure 4.6), consisting of the heavier *b*-quarks, should not be suppressed. The psi′ state is less tightly bound and has a larger radius than the J/psi, so it should disappear first.

The observed J/psi production is currently under intense theoretical and experimental study to see whether it indicates plasma formation. A good cross-check would be to observe that the upsilon production probability is little changed. However, data on upsilon production in heavy-ion collisions will not be available until RHIC runs. Because the upsilon production cross section is very small, a high beam energy is required to create a significant number of upsilons.

TABLE 4.1 Predicted Signatures of Quark-Gluon Plasma Formation

Signature	Process Probed	Experimental Observable
Suppression of J/psi (meson consisting of the bound state of a charm-anticharm quark pair)	Screening of charm-anticharm pair interaction by the surrounding quarks	Decreased probability for J/psi production
Charm enhancement (increase in the total number of charm and anticharm quarks, regardless of whether they are bound)	Fusion of the many gluons produced early in the collision into a charm and anticharm quark pair	Energetic single leptons and pairs of leptons; decays of D-mesons
Thermal photons and lepton pairs	Annihilation of quarks or hadrons with their anti-partners (rate is proportional to the total number of quarks and hadrons and their temperature)	Less energetic pairs of leptons; low-energy photons
Meson mass	Chiral symmetry of phi and rho mesons	Width and decay channel modifications
Strangeness enhancement	Production of additional strange quarks by the plasma	Increased number of mesons and baryons containing strange quarks
Jet energy loss	Nature of the matter encountered by a quark on its way out of the system	Decrease in the production of high-energy jets of particles
Fluctuations in hadron distributions	Bubbles of plasma	Structure in the spatial distribution of hadrons
Time profile of hadron emission	Hadrons emitted slowly from the long-lived mixed phase	Two-particle correlations
Formation of disoriented chiral condensate	Sudden cooling of bubble with restored chiral symmetry	Unusual ratio of charged to neutral pions of very low energy

NOTE: The measurements needed to observe each signature and the physical process underlying the observable are listed.

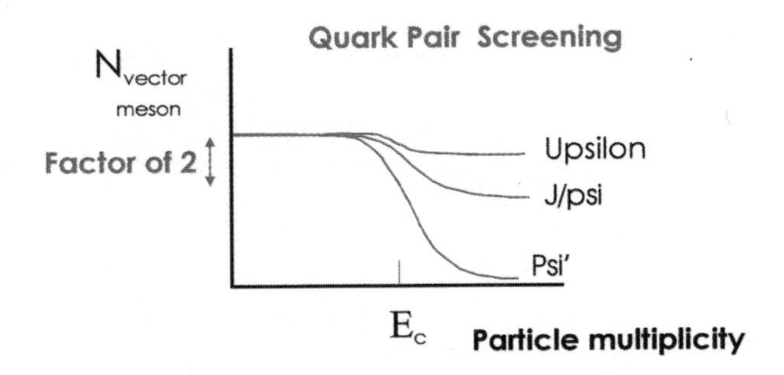

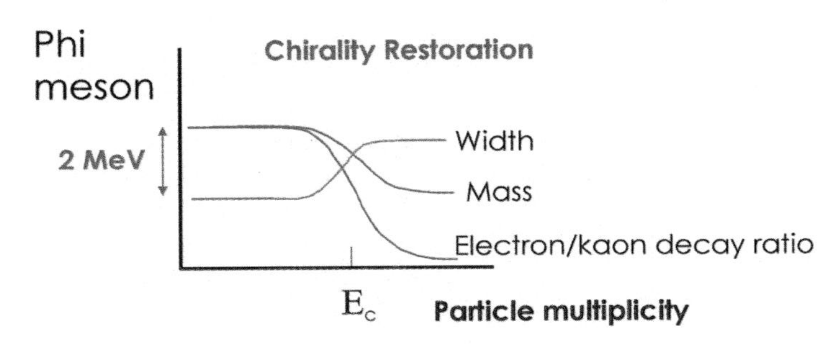

FIGURE 4.5 Quark-gluon plasma and chiral symmetry signatures. This figure illustrates two different signatures of quark-gluon plasma formation and chiral symmetry restoration. The top plot shows the number of J/psi, psi′, and upsilon mesons produced; J/psi and psi′ production should be suppressed if a quark-gluon plasma is formed. The bottom illustrates modification of properties of the phi meson. If chiral symmetry is restored, the phi mass should decrease and width increase; the preferred decay channel should also change. Each is plotted as a function of the energy observed transverse to the beam direction, showing the expected correlated change as the energy density in the collision increases. The magnitude of differences between normal hadronic matter and quark-gluon plasma is indicated.

Copious production of gluons in the early stage of the collision should increase the likelihood for fusion of gluons to produce charmed quarks. This process would lead to an enhancement of charm compared to that in nucleon-nucleon collisions, though only a small fraction of this enhancement goes into the production of charm and anticharm quarks bound together, and this enhancement will not appreciably alter the J/psi suppression discussed above. Mesons incorporating charmed quarks decay before reaching the detectors. Leptons from these decays contribute to the observed lepton pair spectrum in the intermediate mass region labeled 3 in Figure 4.6. Enhanced production of lepton pairs in this region has been observed by two experiments.

Thermal radiation of real or virtual photons (which decay to charged lepton

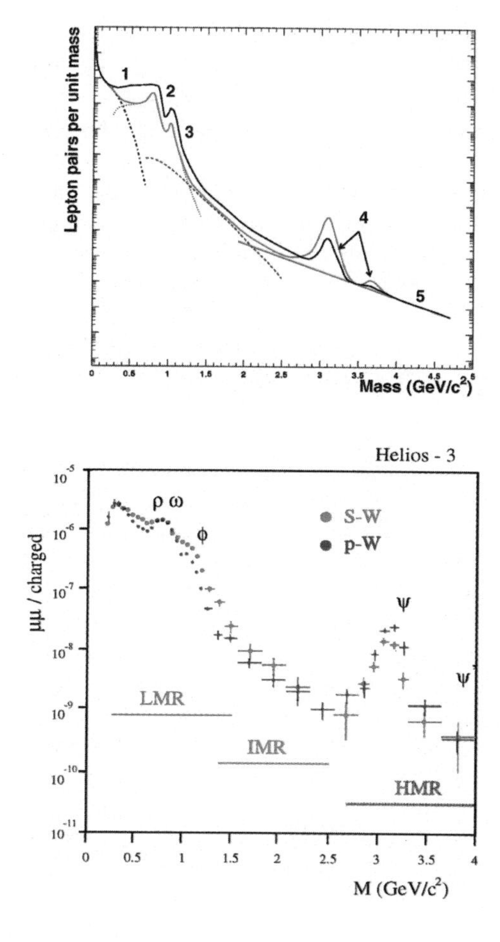

FIGURE 4.6 Lepton pair spectrum. The top panel shows a schematic view of the distribution of lepton pairs (either electrons or muons) as a function of invariant mass. The red line shows the lepton pair spectrum from normal hadronic matter. The dashed blue line indicates the expected spectrum when a quark-gluon plasma is formed. Five regions of the spectrum are of interest: (1) low-mass pairs in which enhanced production from thermal radiation from the plasma should be observable; (2) the phi meson, which should be broader in width as chiral symmetry is restored; (3) intermediate-mass pairs, which should be enhanced by decays of mesons containing charmed quarks formed by the plasma; (4) the J/psi and psi′ mesons, which should be suppressed by the plasma; and (5) the high-mass pairs, Drell-Yan continuum. The bottom panel shows the measured distribution of muon pairs in collisions of sulfur and proton projectiles on a tungsten target. The muon-pair yield relative to the total number of charged particles is given; LMR and IMR indicate the low- and intermediate-pair-mass regions, while HMR shows the start of the high-pair-mass region. Significant differences between sulfur and proton projectiles are visible and may indicate onset of a new phase of matter.

pairs; labeled 1 in Figure 4.6) provides direct information on the thermal history of the collision. Such data are usually studied by looking at the production probability as a function of the total energy or momentum perpendicular to the beam direction carried by the pair of leptons. The highest-energy part of the distribution reflects the hottest, earliest stage of the collision. The mass range in the dilepton spectrum where such radiation appears and the number of dileptons radiated depend on the initial temperature. If very high temperatures are reached, copious thermal dileptons should be produced. Detection of such radiation would determine whether the initial temperature is sufficient for deconfinement.

Excess low-mass dileptons (labeled 2 in Figure 4.6) have been observed in heavy-ion collisions by the CERES and HELIOS-3 experiments. However, attempts to explain the excess as thermal radiation from the hot initial state have not been able to account for its mass dependence. Thermal radiation should decrease smoothly with increasing dilepton mass, but the experiments show a peak directly below the rho meson. This can be explained if the mass of the rho meson were to decrease during the hot, dense stages of the collision. These data have generated considerable excitement, as decreasing rho mass may indicate the onset of chiral symmetry restoration.

In proton-proton collisions, hadrons containing a strange quark are produced much less frequently than hadrons with light quarks only. The high temperature required for a quark-gluon plasma should make it much easier to produce the heavier strange quarks, enhancing strangeness. Indeed, experiments have found a doubling of the number of strange hadrons compared with proton-proton yields. An enhancement in the number of phi mesons, which are bound states of strange and antistrange quark pairs, has also been observed. The excess strange particles fall into two classes. The yields of the lighter strange particles, in particular the kaons and lambdas, can be accounted for by secondary collisions. If a sufficiently energetic pion collides with a nucleon, a strange and antistrange quark may be created (labeled 2 in Figure 4.6). These would get bound into an extra kaon and lambda. Models simulating the cascade of hadronic scatterings correctly reproduce the observed kaon and lambda excess in heavy-ion collisions. However, the other strange particles, phi mesons and strange antibaryons, which contain more than one strange quark or antiquark, are not so easily explained. Strange antibaryons are nearly impossible to produce in secondary collisions among the hadrons in the hadron gas phase of the collision. All of the existing models must invoke new physics (e.g., statistical decay of small quark matter clusters) to explain the experimental results. Consequently, production of strange antibaryons will be studied extensively at RHIC.

A quark-gluon plasma affects particles traversing it differently than does hadronic matter. A single quark, detected after hadronization and freezeout via the jet of particles it produces, should lose significantly more energy going through a deconfined plasma of quarks and gluons than if the quarks and gluons are bound into hadrons. The plasma will also affect the angular correlation between pairs of jets. While the beam energy in experiments performed to date is too low for significant jet production, energy loss of jets is an important new signal to be studied at RHIC.

Unusual objects may be made when a quark-gluon plasma expands and hadronizes. We must remain prepared for nature to surprise us in the way she reveals the physics of this unexplored regime, as when first presenting neutron stars to us in the form of pulsars.

CHIRAL SYMMETRY

At high temperatures or densities, chiral symmetry may hold, and consequently the properties of hadrons may be modified with increasing density of the medium around them. For example, the mass of the kaons is predicted to decrease, an effect which would be manifested experimentally via a larger number of very low momentum kaons. The mass of the rho meson would decrease, modifying the lepton pair spectrum as illustrated in Figure 4.6; this may possibly already have been observed by the CERES collaboration. The mass and width of the phi meson would also be affected. Because the phi is close in mass to two kaons, a slight modification in the kaon mass could cause a measurable change in the likelihood of phi decay to two kaons versus (for instance) decay to an electron plus a positron.

A new signature for the chiral phase transition has been proposed, the formation of disoriented chiral condensates corresponding to different ways of going from a phase with full chiral symmetry back to symmetry of the kind found in normal matter. It is possible that this transition does not happen uniformly over the entire system, resulting in regions with different conditions. Such a state would have highly specific experimental characteristics, such as unusual ratios of charged to neutral pions, or charge correlations, as have in fact been reported in certain cosmic ray events. Searches at currently available energies have not yet produced evidence for such a condensate, but at the much higher energies of RHIC, a larger fraction of the collisions may be expected to reach the required conditions, substantially increasing the probability of observing these signatures.

RELATIVISTIC HEAVY-ION COLLIDER

The heavy-ion collisions studied to date at Brookhaven National Laboratory's AGS and at CERN in Europe show hints of interesting new phenomena. However, we do not yet have firm evidence of the existence of a new phase of matter. RHIC at Brookhaven National Laboratory, on which construction is well under way, is expected to settle these questions. Beginning in 1999, RHIC will collide beams of Au ions at energies 10 times larger, in the center of mass, than have been available at the AGS or at CERN. Construction is proceeding on schedule, and, as of this writing, beam was successfully brought into RHIC from the AGS (which will continue to run, primarily to serve as an injector for RHIC) and brought one-sixth of the way around the RHIC ring. For further information on this facility, see Chapter 8.

Experiments at RHIC

RHIC has six collision regions, four of which will be instrumented for data taking in 1999. The four experiments include two large, multipurpose experiments, STAR and PHENIX, and two small, focused experiments, PHOBOS and BRAHMS. This group of experiments was chosen to be complementary and allow full measurement of possible new phenomena when RHIC begins operation. The experiments were designed to cover the predicted signatures of new physics, as listed in Table 4.1. Complementary experiments are required, because it is not possible to measure all of the signatures in a single experiment; for example, detecting all the hadrons interferes with measurement of low-energy leptons.

STAR is optimized for large coverage for hadrons, and PHENIX for leptons and photons. Each one surrounds the collision area with a variety of detectors and catches as many of the emitted particles as possible. STAR and PHENIX are both international collaborations of 300 to 400 physicists. The two smaller experiments, PHOBOS and BRAHMS, are each approximately one-tenth the size and complexity of the large experiments. PHOBOS will measure the very low energy hadrons that do not penetrate the large magnetic fields and thicker detector layers in the large experiments. BRAHMS will reach angles much closer to the beam direction than any of the other experiments, thus providing a more kinematically complete picture than the other experiments, albeit with only a few particles from each event.

Figure 4.7 shows the layout of the PHENIX experiment at RHIC. The detectors in PHENIX measure as many of the predicted quark-gluon plasma signatures as possible, with an emphasis on leptons and photons. This will allow a search for correlated changes as the beam energy and size of the colliding system is varied.

Construction is currently under way in each of the collision areas in RHIC. All four detectors will be ready to collect data in 1999. Figure 4.8 shows the status of PHENIX construction as of September 1998, with the magnets and detector carriages shown in Figure 4.7 installed into the collision area.

The four experiments under construction for 1999 constitute a first survey of the physics of RHIC. They cover most, but not all, of the predicted signatures of the quark-gluon plasma. Analysis of early data will show which detector capabilities require upgrades to fully realize the physics potential of RHIC. The two uninstrumented beam intersections offer the possibility of entirely new experimental approaches. The two large experiments have both made compromises (driven by financial considerations) in their initial detectors, but upgrade paths to

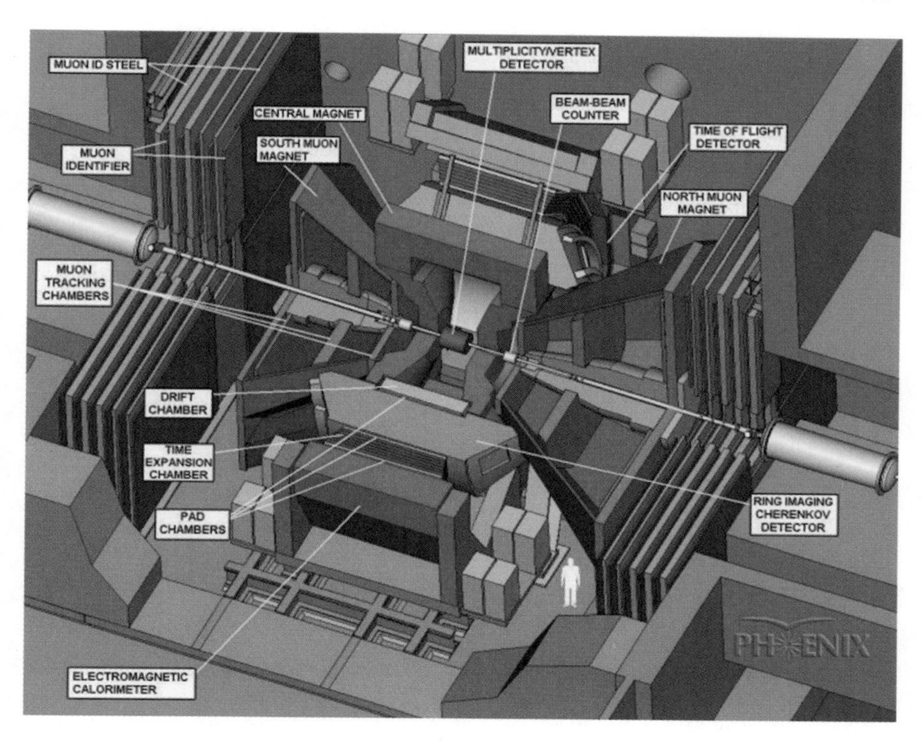

FIGURE 4.7 The PHENIX detector for RHIC. Different components of the detector are labeled; each component provides information complementary to the others. Information from the components is combined to determine the momentum, type, and trajectory of the particles produced in the collision. The two beams enter from either side of the experimental hall and collide in the center of the detector. PHENIX has three magnets; the central magnet analyzes electrons and hadrons, while the forward and backward magnets measure muons emitted near the beam direction. Charged particles entering the central magnet are tracked by using three types of gas-filled detectors. Hadrons are identified by measuring their arrival time at walls of scintillator detectors. Electrons are identified in four ways: production of light in the ring-imaging Cerenkov counter, the distinctive amount of energy lost when they traverse the time expansion chamber, the unique shower of energy produced in the electromagnetic calorimeter, and matching of the shower energy with the momentum measured by the tracking system. Photons, which are neutral, are identified by the energy they deposit in the electromagnetic calorimeter and the absence of a charged track associated with that deposition. Muons are the only particles able to penetrate the central magnet pole face and the absorber layers at the ends of PHENIX. Their momentum is measured with tracking chambers located inside the muon magnets. (Courtesy PHENIX group.)

FIGURE 4.8 The PHENIX collision area in September 1998. The beams enter the area from left and right. In the picture, the central magnet and both muon magnets are installed. The other detector elements will be brought to the hall as they are completed. (Courtesy PHENIX group.)

extend their capabilities have been identified. The data and new discoveries will determine which further studies and upgrades are required.

OUTLOOK

Intriguing experimental results show that the high-energy-density matter created in heavy-ion collisions is at (or maybe just above!) the threshold to enter a new form. RHIC will come online in 1999, with 10 times higher energy than currently available, and will produce matter well over the threshold for the transition to deconfined matter. Theoretical descriptions of the quark-gluon plasma have predicted numerous signatures of its formation, and a suite of experiments is under construction to investigate the proposed signatures. The properties of high-energy-density matter will be extracted, and the data will improve the current understanding of how quarks are confined and how particles acquire mass when the quarks coalesce into hadrons. As we begin to learn the nature of high-energy-density matter at RHIC, the experiments can be accordingly expanded and upgraded. Finally, there is a possibility that complete surprises await. The energy regime is totally new, and the potential for unexpected discoveries is exciting. The answers will begin to be known in the next several years.

5

The Nuclear Physics of the Universe

INTRODUCTION: CHALLENGES FOR THE FIELD

The rich tradition of collaboration between nuclear physics and astrophysics dates from the early work of Hans Bethe and Willy Fowler, nuclear physicists who won Nobel Prizes for their efforts to understand the nuclear reactions taking place in stars. Today this intersection remains especially vital, driven on the one hand by the rapid technological advances in astronomical observation, and on the other by the need to understand the underlying nuclear and atomic microphysics that govern most astrophysical objects and phenomena.

The questions nuclear physicists ask about our universe range from the synthesis of the light elements in the first minutes after creation to the violent deaths of massive stars we observe today:

- What is the explanation for the shortfall of neutrinos emitted by our Sun? Is the current discrepancy due to new particle physics, beyond our standard theory of electroweak interactions, or does it represent some misunderstanding of the nuclear reactions that power our Sun? What new technologies can nuclear physicists and others exploit to measure the entire spectrum of solar neutrinos?
- Is the process by which the lightest elements were created in the first few minutes of the Big Bang well understood? Might there exist, in this fossil record of the birth of the universe, evidence of early exotic states of high-temperature hadronic matter?
- What drives the spectacular stellar explosions known as supernovae? The detection of the various neutrino species emitted in such explosions may deter-

mine whether massive neutrinos play a crucial role in cosmology. What detectors might nuclear physicists construct for this purpose?

- Our Earth and its rich biology depend on the many heavy elements synthesized during stellar evolution and in violent events like supernovae. What are the nuclear processes responsible for nucleosynthesis, and when and where do they take place?
- What exotic forms of nuclear matter exist at the extraordinary densities characteristic of neutron stars? What connections can be established between the observed properties of such stars—their masses, radii, rotation rates, electromagnetic emissions, and so forth—and the behavior of nuclear matter under exotic conditions?
- As the early universe cooled, a hot plasma of unconfined quarks and gluons coalesced into a gas of mesons and nucleons. Can ultrarelativistic heavy-ion collisions provide new insight into the consequences of this phase transition?
- Earth is bathed in a sea of cosmic radiation, much of it emanating from nuclear processes occurring in our galaxy. How can further measurements of nuclear properties—lifetimes, gamma-ray lines, and so forth—help in determining the origin and consequences of this radiation? How can we exploit unstable nuclei as cosmological clocks of past events in our galaxy?

The efforts under way to address these challenges are described below.

THE SOLAR NEUTRINO PROBLEM

Stars, to sustain themselves against the force of gravity, must maintain the pressure of their gases by constantly producing energy. In our Sun, this energy is generated in a series of nuclear fusion reactions in which four hydrogen atoms are converted into helium. These reactions take place deep in the solar core, where temperatures are sufficiently high to allow nuclear fusion to occur. Although we cannot see into the solar core by conventional means, these reactions do produce one form of radiation, neutrinos, to which the Sun is transparent. Passing through the cooler outer layers of the Sun without scattering, these neutrinos carry, in their flux and energy distribution, a detailed record of the reactions by which they were produced. Thus, they offer a unique opportunity to view the nuclear processes that power stars like our Sun.

But the reason neutrinos can pass so easily through the Sun—their remarkably weak interactions with matter—also means that detecting them on Earth is a formidable experimental challenge. After almost three decades of effort, the tools to answer that challenge may be in hand. In the summer of 1965, a group of nuclear scientists began excavations for the first experiment, deep within the Homestake gold mine in Lead, South Dakota (see Box 5.1). With a detector filled with 610 tons of cleaning fluid, the experimentalists patiently waited for rare reactions of neutrinos that would convert a chlorine atom into argon. Be-

BOX 5.1 Solar Neutrinos from Homestake to the Sudbury Neutrino Observatory

In the summer of 1965, workers deep in the Homestake gold mine, Lead, South Dakota, completed the excavation of a $30 \times 60 \times 32$ ft^3 cavity at a depth of 4,850 ft. This was the first step in bringing to life a new detector proposed by Ray Davis, Jr., and his Brookhaven National Laboratory collaborators. The cavity was soon occupied by an enormous tank filled with 610 tons—the contents of 10 railway tankers—of chlorine-bearing cleaning fluid. The purpose of this detector was to make the first attempt to verify that our Sun produces its energy by converting four protons into ^{4}He in a sequence of nuclear reactions called the *pp* chain (Figure 5.1.1). This prediction was the basis of our understanding of stellar evolution. Three years later Davis, Harmer, and Hoffman announced an upper bound on the solar neutrino flux that was a factor of two and one-half below the theoretical expectation. Davis was consoled by one of the miners, who pointed out that it had been, after all, an unusually cloudy summer.

FIGURE 5.1.1 The ^{37}Cl detector located in the Homestake gold mine, Lead, South Dakota. The steel vessel contains 0.6 kilotons of percloroethylene. During use, the cavity is filled with water to provide additional shielding from neutrons and other radioactivity produced in the surrounding rock walls. (Courtesy of Brookhaven National Laboratory.)

The chlorine detector was literally 20 years before its time. It exploited a marvelous circumstance: solar neutrinos would convert a few atoms of ^{37}Cl into ^{37}Ar which, because argon is a noble gas, could be quantitatively flushed from a large volume of fluid by a helium-gas purge. As the ^{37}Ar then decays back to ^{37}Cl with a half-life of about one month, the number of purged atoms could then, with patience, be counted in miniaturized gas proportional counters capable of detecting these decays. The sensitivity of the technique was astounding: a solar neutrino capture rate producing three ^{37}Ar atoms per week was eventually measured to an accuracy of 10 percent. It would take the community another two decades to develop the technology to mount other experiments capable of verifying the chlorine detector results and of extending its measurements to other portions of the solar neutrino spectrum.

Today, with the completion of the Kamioka experiment and the SAGE and GALLEX measurements of the low-energy solar neutrino flux, it is recognized that the solar neutrino problem is a profound issue for all of physics. It has become increasingly difficult to account for the discrepancies by plausible changes in the solar model or in nuclear reaction rates. On the other hand, the results appear consistent with new particle physics, particularly the possibility that the electron neutrino has a mass and oscillates with either the muon or tauon neutrinos. If this is the correct explanation, then massive neutrinos would contribute in cosmology to dark matter and to the clustering of visible matter on large scales. In particle physics, a new Standard Model would have to be formulated, one in which massive neutrinos are accommodated.

Today, two major new solar neutrino experiments have begun with the goal of providing an unequivocal resolution to the solar neutrino problem. The first is SuperKamiokande, a massive 50,000-ton water Cerenkov detector located in the Japanese mine Kamioka. The second is the Sudbury Neutrino Observatory (SNO), the first detector having the capability to distinguish electron neutrinos from muon or tauon neutrinos (Figure 5.1.2).

Just as in the case of the Homestake detector, it is essential to mount these new experiments deep underground so that the Earth above will serve as a protective shield against cosmic-ray backgrounds. The SNO detector, within the Creighton #9 nickel mine at Sudbury, Ontario, Canada, is more than 2 kilometers beneath the surface. The central region of the detector is an acrylic vessel containing 1 kiloton of heavy water, D_2O. Light produced in the detector by neutrino interactions will activate some of the surrounding 9,500 photomultiplier tubes.

The detector depth all but eliminates cosmic-ray interactions, but the confusing effects of natural radioactivity remain a serious problem. The experimentalists had to construct SNO with extraordinarily pure materials. For example, there must be fewer than 10 atoms of uranium or thorium for every million billion atoms of water. It is a great nuclear chemistry challenge to even measure trace contaminations at this level. To ensure such purity, SNO was built under the most rigorous clean room standards, a daunting task in an active mine. If a thimbleful of dust were to enter the 10-story-high detector cavity, the experiment could fail. Finally, the inner heavy-water portion of the detector is protected by a 7-kiloton shield of ordinary water, which will absorb neutrons produced by radioactivity in the surrounding rock walls.

Why does the inner detector consist of rare heavy water? This has to do with proving that neutrinos oscillate, an effect that can be especially pronounced for solar neutrinos due to the matter effects discussed in the text. One kind of reaction

BOX 5.1 Continued

FIGURE 5.1.2 The Sudbury Neutrino Observatory, viewed through a fisheye lens, just before completion. The inner spherical acrylic vessel will isolate SNO's 1,000 tons of heavy water from the surrounding ordinary water that will fill the entire cavity when the detector is operating. The concentric darker structure is the array of 9,500 photomultiplier tubes (PMTs) nearing completion. The PMTs view light produced in the inner water volumes and are temporarily covered with removable shields to maintain low levels of surface contamination. The PMT array is 18 m in diameter. The detector is located 2,000 m underground in the INCO mine in Sudbury, Ontario, Canada. (Courtesy of Lawrence Berkeley National Laboratory.)

occurring in heavy water—neutrino absorption on deuterium to produce two protons and an electron—can only be initiated by the electron neutrinos, and thus will count the electron neutrino portion of the solar neutrino flux. But a second reaction—neutrino scattering off deuterium to produce a neutron and a proton—will count all of the solar neutrinos. Thus, if these two reactions indicate different fluxes, we will know that oscillations have produced neutrinos that are not of the electron type.

As of the writing of this report, SuperKamiokande has reported about a year of data. The construction of the SNO detector has just been completed and commissioning is under way, with first results expected some time in the next year. Thus, the mystery story that began with Ray Davis's experiment three decades ago is approaching its exciting climax.

cause argon is a noble gas, the few atoms produced, roughly one every two days, could be extracted from the Homestake detector and counted by observing their subsequent decay. This experiment, which continues to operate and improve in accuracy, produced the first evidence that the Sun was not producing as many neutrinos as predicted by our theories of stars and nuclear reactions. Because of this intriguing result, an experiment of a different kind, in which neutrino reactions could be measured event by event in a detector containing 4,500 tons of ultrapure water (neutrinos produce detectable light when they scatter off the electrons in water), was mounted in the Kamioka mine in Japan. The Japanese physicists and their American collaborators again found a deficit of neutrinos. Finally, two radiochemical detectors that use gallium, a special material sensitive to the lowest-energy neutrinos produced by the Sun, were constructed in Italy and Russia. U.S. nuclear physicists played important roles in both of these efforts.

The combined results of these measurements are surprising. The flux of the highest-energy neutrinos produced by the Sun, coming from the beta decay of ^{8}B, is about 40 percent of the standard solar model value, while those produced from electron capture on ^{7}Be are reduced even further. While a lower neutrino flux suggests that the Sun's core might be cooler than predicted, such a change is expected to suppress the ^{8}B neutrino flux much more than the ^{7}Be flux, in contradiction to the measurements. In fact, detailed fits to the experimental results show that even with arbitrary changes in the total fluxes of the ^{7}Be, ^{8}B, and very low energy *pp* neutrinos, one cannot fit the data well (Figure 5.1).

The solar neutrino results become more curious in light of other tests that the best current model of the Sun has passed. In recent years, careful measurements of the Sun's surface have allowed astrophysicists to deduce the frequencies of the Sun's internal vibrations, which depend sensitively on the Sun's composition and temperature profiles. The results of such helioseismology studies are in excellent agreement with solar model predictions. In addition, the consistency of stellar evolution calculations with observations of other solar-like stars has increased our confidence in the solar model.

However, the failure to detect the predicted number of neutrinos might have nothing to do with the solar model and nuclear reactions; instead, it could reflect a lack of understanding of the properties of neutrinos. In particular, if neutrinos have a small mass—a possibility not envisioned in the current Standard Model of particle physics—a natural explanation can be offered for the observations. Electron neutrinos produced by the nuclear reaction in stars can then transform (or oscillate) into neutrinos of a different flavor, thereby escaping detection on Earth. Because solar neutrinos are low in energy and travel a great distance before they are detected on Earth, this effect can arise for masses much smaller (e.g., 10^{-6} eV) than those detectable by any other means. Furthermore, it was shown in 1985 that as neutrinos make their way from the solar core through the outer layers of the Sun, oscillation effects can be greatly magnified. This phenomenon is known as the MSW or Mikheyev-Smirnov-Wolfenstein mechanism, after the physicists

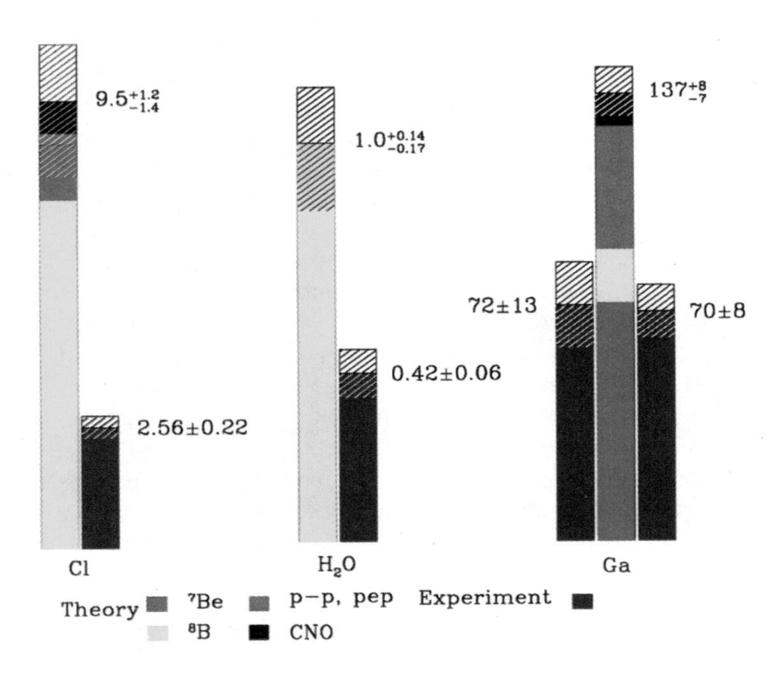

FIGURE 5.1 Comparison of the predicted counting rates of the solar model (taller bars) with the rates measured in the chlorine, Kamioka II/III, and SAGE and GALLEX gallium experiments. The respective uncertainties are indicated by the hashed regions, while the contributions of various neutrino sources to the predicted counting rates are coded by the indicated colors. The counting rates are given in solar neutrino units (SNUs), corresponding to one capture per 10^{36} target atoms per second. The discrepancies between theory and experiment constitute the solar neutrino problem.

who first described it. The dependence of this enhancement on the energy of the neutrino provides a natural explanation for the experimental results discussed above.

If this conclusion is correct, solar neutrino experiments—begun by nuclear physicists to study the nuclear reactions occurring in the solar core—will have provided the first evidence that the Standard Model of particle physics is incomplete. For this reason, there is great excitement about two new experiments now under way. Within the Kamioka mine, a new, purer, and much larger water Cerenkov detector has recently been completed. It will provide a more precise measurement of the shape of the ^{8}B neutrino spectrum, perhaps revealing distortions characteristic of neutrino oscillations. Named SuperKamiokande, this detector was built by a Japanese-U.S. collaboration (Figure 5.2). A second detector, a Canadian-U.S.-U.K. collaboration, the Sudbury Neutrino Observatory (SNO), is similar in design, except that the water is "heavy," with the hydrogen replaced by deuterium. This difference will allow the nuclear physicists building the detector to measure solar neutrinos whether or not they have oscillated into a different type (or flavor). If this detector measures more neutrinos than seen in previous experiments, it will prove that neutrinos have oscillated, and thus have a nonzero mass. SNO construction is finished, and the commissioning of the detector is under way. Two other detectors with very low energy thresholds, Borexino and Iodine, are in earlier stages of preparation; when completed, they will provide additional information on the flux and flavor of the ^{7}Be neutrinos.

FIGURE 5.2 The 50-kiloton SuperKamiokande detector was filled in the spring of 1996. Here, members of the collaboration, in a small rubber boat, make use of an opportunity to complete a final cleaning of the phototube surfaces. The inner detector is viewed by 11,200 50-cm-diameter tubes. (Courtesy of the Institute for Cosmic-Ray Research, University of Tokyo.)

The resolution of the solar neutrino problem is important not only to nuclear physics, but also to particle physics, astrophysics, and cosmology. In many theories that attempt to generalize our current Standard Model of particle physics, small neutrino masses are related to entirely new physics that is otherwise beyond our grasp: current and planned accelerators fall far short of the energies where this physics can be seen directly. Thus, neutrinos may provide our only means of experimentally testing these new theories. In cosmology, neutrinos are a leading candidate to explain the problem of dark matter: if the universe is filled with a sea of invisible but massive neutrinos created in the Big Bang, this could explain why the visible matter is insufficient to account for the gravitational forces within clusters of galaxies. Neutrino oscillations could also have profound consequences for supernovae, the violent explosions by which massive stars die once they have consumed all of their nuclear fuel.

THE BIG BANG, THE QUARK-GLUON PLASMA, AND THE ORIGIN OF THE ELEMENTS

We believe our universe began in an extraordinary explosion known as the Big Bang. In the first instants after this explosion, when the temperature was in excess of about 150 MeV, matter existed in a state unknown to us, a plasma of unconfined quarks and gluons. After the universe expanded and cooled below this critical temperature, hadronic matter coalesced into a more familiar gas of nucleons and mesons. It is quite possible that this transition, if sufficiently violent, could have left a signature in our low-temperature world, perhaps through inhomogeneities in the density or by producing isolated regions carrying unusual quantum numbers, such as nuggets of strange quarks. Such relics might be found in today's universe, or we might be able to discern the effects of this transition in the pattern of light nuclei that coalesced out of the inhomogeneous nucleon gas, as discussed below. One exciting nuclear physics connection to these early times is the possibility that future relativistic heavy-ion experiments will allow us to examine this transition in the laboratory, thereby providing the information we need to model the analogous cosmological epoch.

The primordial nucleon soup is the starting point for all the chemical elements that we see about us. In the first few minutes after the Big Bang, the nucleon gas cooled sufficiently to allow protons and neutrons to combine into deuterium, and for subsequent nuclear reactions on deuterium to produce light elements like helium and lithium. Much later, these light elements begin to coalesce under the action of gravity, forming stars and galaxies. Within these stars synthesis continues as nuclei are fused to produce heavier elements and heat: many of the most plentiful elements, like ^{12}C and ^{16}O, are produced in this way. Additional synthesis occurs through violent stellar explosions known as novae and supernovae.

Novae, supernovae, and the "winds" generated by ordinary stars eject the products of nucleosynthesis into the interstellar medium, where they again form the raw material for a future generation of stars and planets. Indeed, were it not for this continuing synthesis, we would lack the essential elements that make up our Earth and allow it to sustain life. One of the primary goals of nuclear astrophysics is to understand quantitatively the processes by which this synthesis takes place.

The synthesis of light nuclei in the very early universe is considered one of the cornerstones of modern cosmology. The abundance and distribution of such nuclei provide a detailed fossil record of conditions in the first few minutes of the Big Bang. By precisely measuring the nuclear reactions that contributed to this synthesis, nuclear physicists have allowed this record to be read with great clarity.

Hydrogen and helium stand out among the elements as being by far the most abundant. While some helium is produced in stars like our Sun, one can demonstrate, by studying the abundances of elements in old stars, that most of the observed helium was produced prior to the formation of stars. Likewise, deuterium, ^{3}He, and lithium appear to be at least partially primordial. These observations are quite consistent with light-element reaction rates measured in the laboratory. Likewise, the absence of stable nuclei with mass numbers 5 and 8 creates roadblocks to further synthesis, accounting for the lack of primordial elements above mass 8. However, the successful prediction of primordial abundances, a crucial result supporting the Big Bang theory, requires one to pick the initial abundance of baryons (protons and neutrons) appropriately.

The baryon density thus derived from Big Bang nucleosynthesis exceeds the density of luminous matter (e.g., stars) we see about us. This problem of the missing baryons, as well as the dark matter problem mentioned earlier in connection with massive neutrinos, prompted nuclear physicists and astrophysicists to explore nucleosynthesis in nonstandard cosmologies, such as those where the matter distribution is quite inhomogeneous due to some early, violent phase transition. The nuclear networks can operate quite differently in an inhomogeneous Big Bang, altering the abundances of elements like Li substantially. The conclusion, after considerable work on the theory and measurements of possible new reaction paths, is that primordial abundances remain a stringent constraint, making it difficult to solve the dark matter problem by altering Big Bang cosmology. Yet, because inhomogeneities could play a more modest role in Big Bang nucleosynthesis, the existence of primordial elements other than those made according to the standard Big Bang model remains an important question.

Progress continues to be made in observations of primordial elements. Deuterium abundances deduced from the absorption of light as it passes through clouds of interstellar hydrogen gas have been at the center of recent Big Bang nucleosynthesis controversies. This is a promising technique that makes use of the power of the newest telescopes. As primordial elements can be produced and destroyed in stars, our increasing knowledge of stellar abundances is also helping us deduce more accurate primordial values from present-day abundance measurements.

In the later synthesis that begins with the formation of the first stars, there is a close coupling between the macroscopic astrophysics issues, models of stars and of galactic chemical evolution, and the nuclear microphysics, the networks of nuclear reactions that govern this evolution. A classic example is the synthesis of ^{12}C from three ^{4}He nuclei, which triggers the ignition of the dense He core of a red giant star. The detailed nuclear structure necessary for this reaction to take place was anticipated because of the astrophysics, then confirmed by nuclear experiments. A reaction of great current interest is the synthesis of ^{16}O from the reaction of ^{4}He with ^{12}C, which determines the relative sizes of the carbon and oxygen shells of massive stars that later explode in supernovae. The sizes of these shells are a crucial factor in predicting the nucleosynthesis that occurs during the explosion. Nuclear physicists are currently trying to measure the rate of this reaction with sufficient accuracy to constrain the astrophysical models.

The expulsion by supernovae of newly synthesized nuclei—both the products of nuclear fusion (common elements like ^{4}He, ^{12}C, ^{16}O, ^{20}Ne, and others) and the many less-abundant species made by the explosion itself—accounts for much of the chemical diversity of our galaxy. For example, about half of the elements heavier than iron must be synthesized in about a second in the rapid neutron capture process, the *r*-process, requiring very high neutron densities and temperatures above a billion degrees. The astrophysical site where such explosive conditions might exist has been a matter of continuing speculation for several decades. One suggestion has been the neutron-rich matter ejected from near the surface of the neutron star that forms at the center of a core-collapse supernova.

An important advance of the past 5 years has been the modeling of the network of nuclear reactions that govern *r*-process synthesis of nuclei in a supernova. One prediction of the theory is that the *r*-process would have occurred in the earliest stars in much the same way as in today's stars, a result in agreement with observations made with the Hubble Space Telescope (HST). Nuclear calculations have recently provided additional evidence that supernovae are the site of the *r*-process: neutrinos produced in such supernova explosions can transmute *r*-process elements into new species. Such neutrino "fingerprints" of a supernova *r*-process site were recently found.

Nuclear physicists are being helped in their efforts to understand nucleosynthesis by the rapid instrumentation advances in astronomy and astrophysics. For example, the origin of the relatively rare elements Li, Be, and B is an important issue for several reasons, including Big Bang nucleosynthesis. The textbook explanation for these elements is that they form when high-energy cosmic rays break up ^{12}C and ^{16}O in the interstellar medium. This mechanism requires preexisting C and O, so it is secondary and should grow in effectiveness as the amount of these heavier elements in the galaxy increases. But recent HST spectroscopic studies of very metal poor stars appear to establish that B grows linearly with Fe and other metals, a signature of a primary process not dependent on preexisting metals. This has focused attention on nucleosynthesis by supernova neutrinos, a process suggested by nuclear physicists: energetic neutrinos scatter off nuclei in the outer layers of the exploding star, knocking out nucleons and transmuting the target nuclei into different elements. Thus ^{11}B can be made by neutrinos scattering off carbon, one of the most abundant elements in the mantles of such stars. A competing possibility is similar spallation reactions of cosmic rays with nuclei in the local environments of stars or nebulae. Anticipated HST measurements separating ^{10}B from ^{11}B may decide which of these processes is more important.

THE SUPERNOVA MECHANISM

One of the outstanding theoretical challenges in nuclear astrophysics is to understand the process by which a massive, fully evolved star ejects its mantle while its core collapses to a neutron star or black hole. The collapse of the iron core to several times nuclear-matter density produces a powerful shock wave that travels outward through the mantle of the star. It is believed that this shock wave, aided by the heating of the matter by neutrinos emitted by the newly formed neutron star, is responsible for the ejection of the mantle. This broad-brush picture has emerged from heroic efforts over a decade or more to simulate this process numerically; difficult hydrodynamics, the need to understand the nuclear equation of state at high densities and large neutron number, shock wave propagation, general relativistic effects, and the complexity of the neutrino diffusion make this problem unusually challenging. Despite great progress, the field is still struggling to define a mechanism that is sufficiently robust to account for the observed frequency of supernovae.

An exciting recent advance has been the development of two-dimensional hydrodynamic codes for simulating stellar collapse. This step, made possible by supercomputers, is important because observations of Supernova 1987A (Figure 5.3), the most recent supernova in our galactic neighborhood, provide clear evidence of deep mixing that can only be generated dynamically in multidimensional models. Recent calculations find convective cells forming above the neutrinosphere (the point in the star where neutrinos are no longer "trapped," but can stream out freely, as they do in less dense stars like our Sun). These cells sweep colder matter to smaller radii, where it can be more effectively heated by neutrinos. Hot material is carried to the shock by buoyancy; once there, it can help to push the shock outward. One group of supernova modelers finds that the enhanced neutrino heating produces a successful explosion for most initial conditions. A competing group finds less favorable results. The discrepancy appears to reflect different approximations in the treatment of neutrino diffusion and of the innermost layers of the star, problems that can be resolved if supercomputer resources are increased.

A result that is independent of the details of the explosion is the characteris-

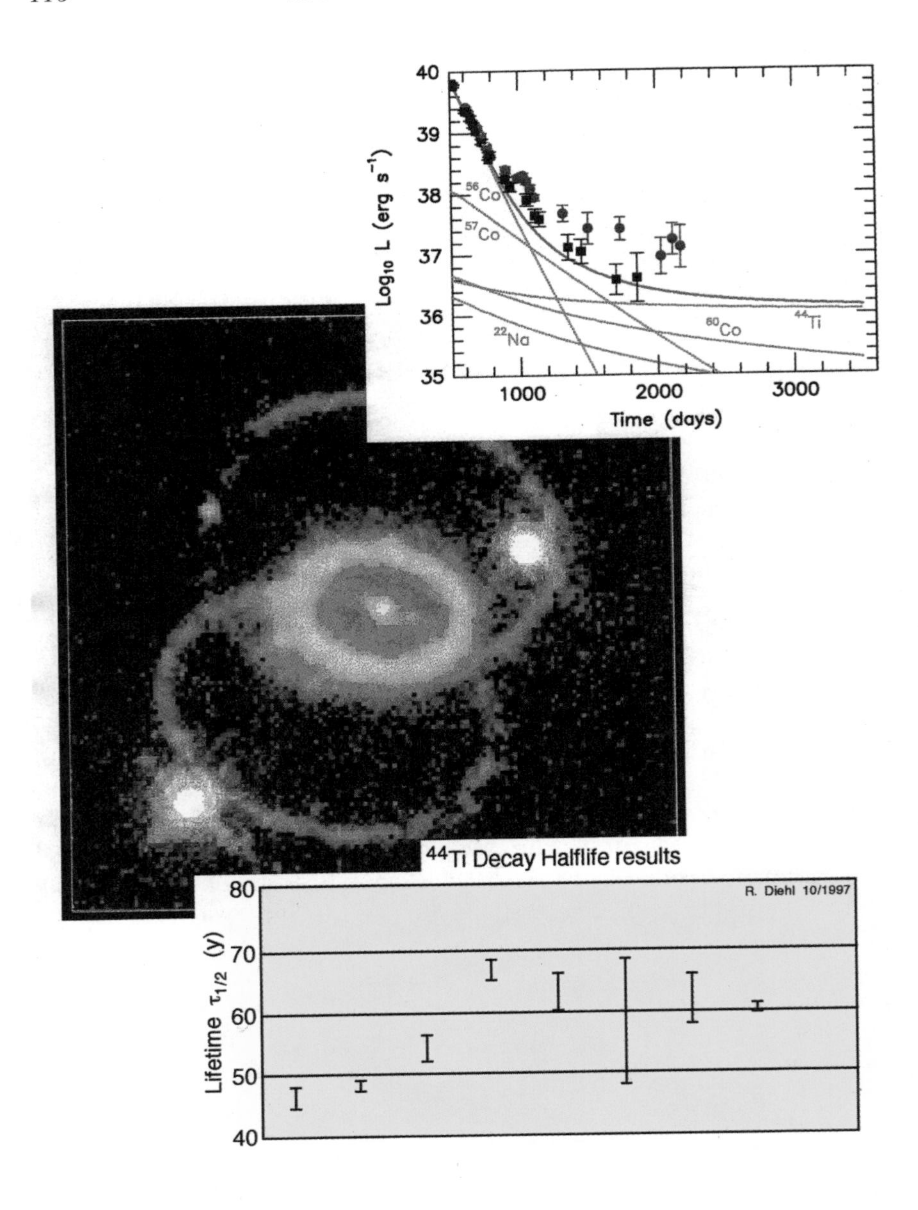

40
39
38
37
36
35
Log₁₀ L (erg s⁻¹)
56Co
57Co
22Na
60Co
44Ti
1000
2000
3000
Time (days)
44Ti Decay Halflife results
R. Diehl 10/1997
80
70
60
50
40
Lifetime τ1/2 (y)

tic hierarchy of neutrino temperatures: the muon and tauon neutrinos are about twice as energetic, on average, as the electron neutrinos. It has been emphasized recently that, if the tauon neutrino has a mass large enough to be important in cosmology, oscillations between electron and tauon neutrinos might be expected in supernova explosions. This would lead to a temperature inversion between electron and tauon neutrinos, producing interesting effects in the explosion mechanism and in the nucleosynthesis from neutrino-heated supernova ejecta, e.g., destroying the conditions necessary for the *r*-process. More important, by measuring the types and energy distribution of the neutrinos emitted by a galactic supernova, experimentalists should be able to deduce whether the tauon neutrino is the source of the nonbaryonic dark matter: the temperature inversion will cause a dramatic increase in electron neutrino capture rates. Several of the solar neutrino detectors that are being built or designed by the nuclear physics community are motivated in part by this additional goal. Operation of such detectors for extended periods is necessary to provide a reasonable probability of detecting these rare events in our galaxy.

The success or failure of the supernova mechanism in a star of given mass is critically dependent upon the properties of the iron core. The structure of the core depends both on the nuclear reactions governing the evolution of the star prior to collapse and on weak interactions occurring in the early stages of the collapse: electron capture on nuclei produces neutrinos that immediately escape the star, thus lowering the number of electrons and electron neutrinos trapped in the core. The electron capture proceeds by spin-flip transitions to states in the daughter nucleus. One of the important experimental contributions to the supernova problem has been careful measurements of spin-flip strengths in nuclei, using (p,n)

FIGURE 5.3 Hubble Space Telescope picture of Supernova 1987A, which exploded 169,000 years ago in a nearby dwarf galaxy, the Large Magellanic Cloud. The light from this explosion reached Earth in 1987 and was visible to the unaided eye in the southern hemisphere. Three rings appear around the explosion region. While their origin is not precisely understood, they may arise from irradiation of a bubble of gas blown into space by the supernova's progenitor star. The supernova ejecta include many newly synthesized nuclei. The presence of some of these can be deduced from the light produced by the supernova as a function of time, shown at the upper right. The observations are consistent with a gamma-ray power source from the decay of short-lived nuclei such as ^{56}Co, ^{57}Co, and ^{44}Ti. Note that the light curve at late times depends on ^{44}Ti, whose half-life was quite uncertain until recently. Nuclear physicists have made several new and independent measurements, shown in the figure at the bottom, that fix its half-life at about 60 years, so that ^{44}Ti should continue to dominate the SN1987A light curve for the next century. (Courtesy Space Telescope Science Institute.)

and (n,p) reactions at medium-energy nuclear facilities, as described in the next section.

During a supernova collapse, the matter at the core of the star reaches the extraordinary densities characteristic of the nucleus (a million billion times that of ordinary matter). At this point, as individual nucleons begin to touch, the strong nuclear force halts the collapse. The result is a trampoline-like rebound that sends a shock wave traveling outward through the core (Figure 5.4). The dynamics of this process—compression of the matter to and past nuclear density, rebound, and formation of the shock wave—is governed by the nuclear equation of state, a topic discussed in more detail in the neutron star subsection.

MEASURING STELLAR NUCLEAR REACTIONS IN THE LABORATORY

Careful treatments of the energy dependence of nuclear reaction rates are required to extrapolate higher-energy measurements to characteristically low astrophysical energies. Similarly, many explosive stellar reactions involve exotic isotopes not found naturally on Earth. Here, laboratory nuclear astrophysics is making crucial contributions to our understanding of stellar structure.

In the solar neutrino problem, the primary remaining nuclear uncertainty is the rate for proton capture by ^{7}Be, a reaction that leads to the production of high-energy ^{8}B solar neutrinos. These are the neutrinos that react in the SNO and SuperKamiokande detectors. Two low-energy data sets agree in energy dependence, but disagree in normalization by about 25 percent. Theory confirms that the energy dependence is correct, but it probably is not yet capable of predicting the absolute cross section. Several new experiments to resolve this problem are now being pursued. These include both conventional measurements, in which a ^{7}Be target is bombarded with protons, and new approaches with different systematic uncertainties that will be possible at radioactive beam facilities. For example, one could use a hydrogen target and a ^{7}Be beam, or observe the inverse process, breakup of ^{8}B in the Coulomb field of a heavy nucleus. It appears likely that uncertainties in this rate will soon be reduced substantially.

Nuclear theorists have done much of the work on atomic screening corrections that must be made to derive stellar cross sections from terrestrial measurements. In the solar core, almost all of the electrons in the matter are free, and not confined to orbits around the nuclei as they are in terrestrial atoms. Thus, theorists must extract from laboratory cross sections values that are appropriate for the bare nuclei found in stars.

Neutrino physics is crucial to supernova explosions, controlling the transition from ordinary to neutron star matter and helping to power the explosion itself. Nuclear accelerators producing medium-energy protons have made a major contribution to our understanding of these processes. When a proton scatters off a nucleus, producing an outgoing neutron in the forward direction, it probes

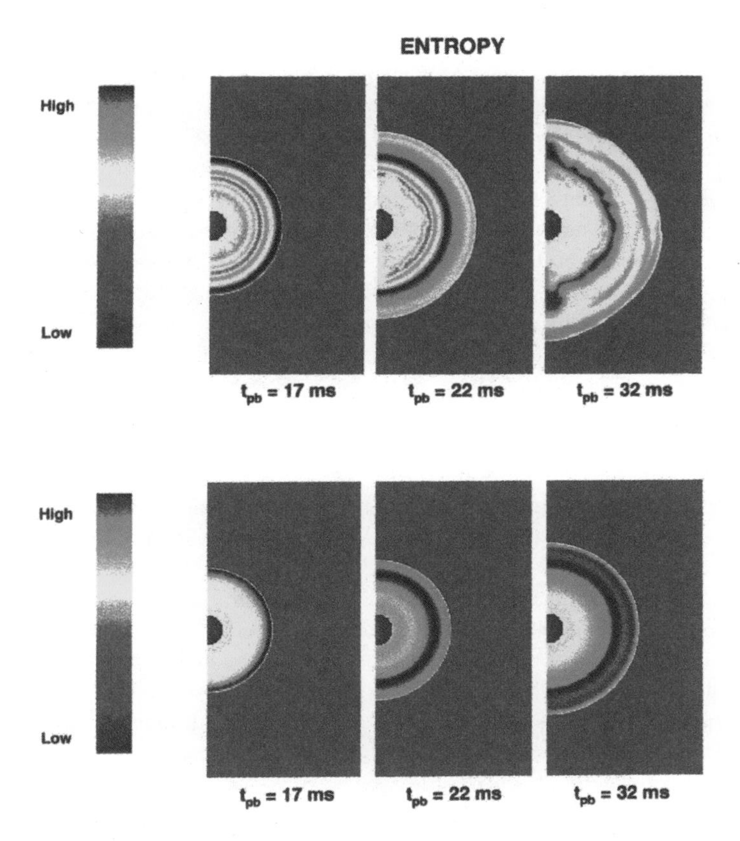

FIGURE 5.4 This series of two-dimensional plots shows the results of a supercomputer study of a supernova explosion in a 15-solar-mass star. The frames correspond to "snapshots" of the entropy, or degree of disorder, in the material above the neutron star's surface some tens of milliseconds after core bounce (the time when the collapsing star reaches maximum nuclear density). The upper frames show a simulation without neutrino transport, while the lower frames have neutrino transport included. The higher regularity of the lower frames indicates that neutrino transport tends to inhibit the development of convection in such two-dimensional simulations. The calculations were performed by using facilities at several supercomputer centers.

the same spin-flip transitions that govern neutrino-nucleus interactions. (The same measurements can be made for radioactive species of interest by illuminating a hydrogen target at a radioactive ion beam facility.) Thus, such (p, n) and (n, p) reactions can determine the beta decay rates important during stellar collapse, as well as the neutrino-nucleus reaction rates that help to eject the outer layers of the star. This technique has also been important in calibrating solar neutrino detectors. In fact, it led to the discovery that the original cross section estimates for the chlorine detector were based on a flawed assumption; fortunately, it proved possible to correct the error.

The nuclear reactions that make up the *r*-process involve exotic, short-lived, neutron-rich isotopes that can exist only in explosive conditions similar to those in a supernova. Many of these isotopes cannot yet be produced on Earth. To understand the unusual properties of the nuclei along the *r*-process path, nuclear experimentalists are building radioactive ion beam facilities capable of producing very short-lived nuclei, as mentioned in Chapter 3. Such facilities will provide direct information on selected important nuclei on the *r*-process path, and they will determine systematic trends for neutron-rich nuclei that will lead to more accurate theoretical predictions of the masses, lifetimes, and decay properties of other *r*-process nuclei.

Radioactive beams will help us understand another kind of explosive event important in novae and x-ray bursters. Novae are believed to occur in systems containing two stars, one a white dwarf that has completed its nuclear burning and the other a normal star, that is still evolving. Material from the evolving star can be accreted onto the white dwarf, where it ignites in explosive reactions that convert hydrogen to heavier elements. The elements observed in the ejecta of novae serve as a diagnostic of the conversion process. Most of the star's energy is believed to come from the conversion of hydrogen into helium through a series of reactions called the CNO cycles, so named because carbon, nitrogen, and oxygen act as catalysts for the reactions. But if sufficiently high temperatures are reached, as in the case of an x-ray burster where the accreting object is a neutron star, the reactions break out of the CNO cycles, producing heavier nuclei by repeated capture of protons or helium nuclei, with intervening radioactive decay (Figure 5.5). This rapid proton capture, or *rp*, process is characterized by much faster energy release, accompanied by the synthesis of distinctive proton-rich nuclei. These nuclei can serve as sensitive probes of the stellar conditions during novae, provided the nuclear reactions responsible for their creation, including their dependence on stellar temperature and densities, are well determined. At present, the *rp* process is understood only in broad outline. For example, the conditions for breakout to the *rp* process are poorly known, as are the effects of various nuclear reaction bottlenecks that limit the flow toward higher masses. Such bottlenecks determine the time scale for the *rp* process and thus the necessary characteristics of the associated stellar explosion.

This situation should improve greatly with the coming new generation of radioactive ion beams. The rates for particle reactions on ^{14}O and ^{15}O, crucial in determining the temperatures and densities where breakout from the hot CNO cycle to the *rp* process occurs, can be directly measured with planned radioactive beams (though the beam intensities and detector requirements are challenging). Such facilities will also allow experimentalists to measure the masses and disintegration energies of *rp* process bottleneck nuclei near the proton drip line, such

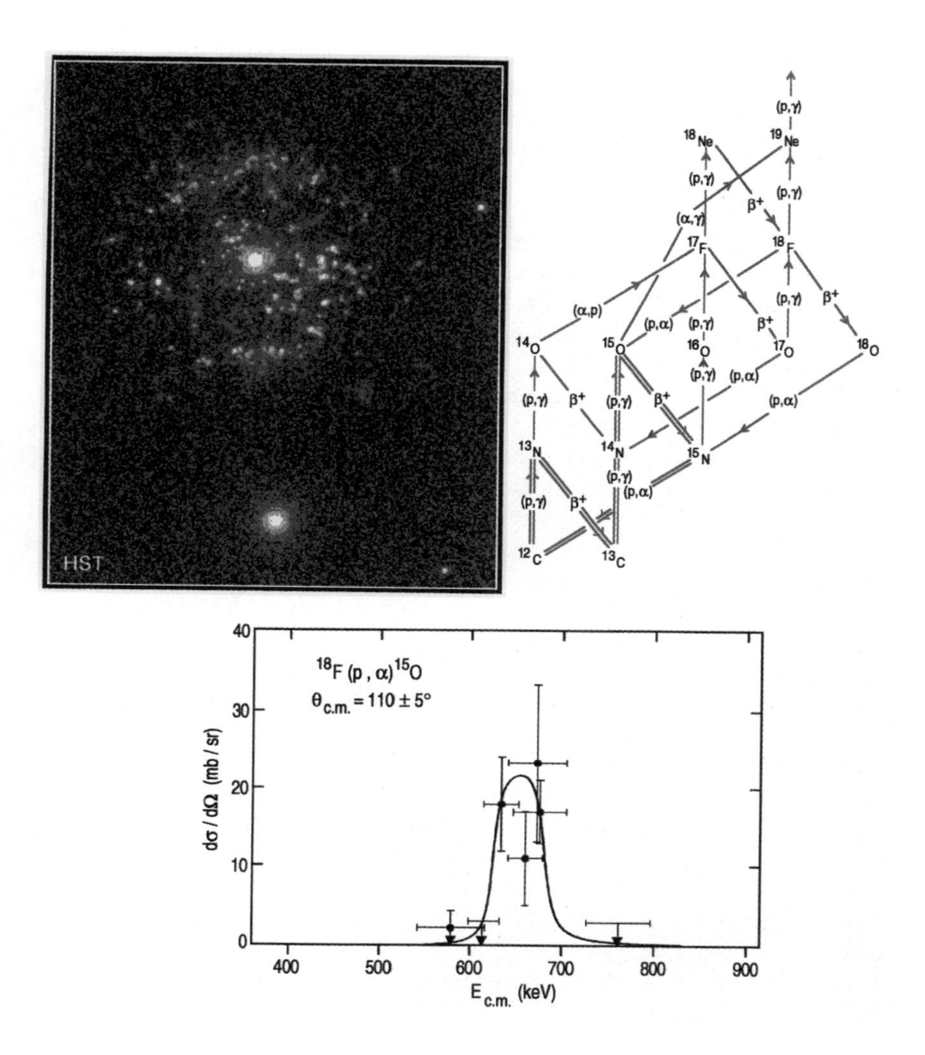

FIGURE 5.5 At the upper right, the nuclear reaction network of the carbon-nitrogen-oxygen (CNO) cycles, including possible pathways for breaking out of these cycles into heavier nuclei, is shown. Some of the branching points involve reactions on short-lived nuclei. At the bottom, the results from a recent measurement of one of the relevant pathways, $^{18}F(p, \alpha)^{15}O$, is shown. ^{18}F has a half-life of about two hours. Above is a Hubble Space Telescope photograph of the unusual recurrent nova, T Pyxidis, which erupts about every 20 years. Blobs of gas, spread over an area about one light-year in diameter, are the "shrapnel" from previous explosions. The thermonuclear runaways that power typical novae release an energy equivalent to about 10^{20} tons of TNT. (Courtesy Space Telescope Science Institute.)

as ^{68}Se. Our current poor knowledge of the disintegration energy of this nucleus makes the stellar lifetime of ^{68}Se uncertain by about a factor of 10,000. Such experimental studies have the potential to determine the rate, principal nucleosynthetic products, and endpoint of the *rp* process (Figure 5.6). This, in turn, would make the *rp* process an important probe of x-ray bursters and other explosive stellar environments.

Additional nuclear physics tools are becoming increasingly important for probing stars and the interstellar medium, as a result of recent instrumentation advances in satellite-based gamma-ray observations. Gamma-ray lines from excited states in ^{12}C and ^{16}O, detected from the direction of the Orion nebula, indicate that such nebulae could synthesize light elements like lithium, beryllium, and boron by cosmic-ray spallation; gamma-ray lines from ^{26}Al decay have been used to map the galactic distribution of this element, a result that is helping to eliminate some of the proposed sites for Al synthesis. Radioactive nuclei and the gamma-ray lines they produce are a powerful probe of our galaxy because, through their varying lifetimes, one can sample different epochs in time.

Such measurements also provide crucial information about the nature of supernova progenitors and supernova explosions. Much of the light emitted from supernova remnants is associated with decays of isotopes like ^{44}Ti, ^{56}Co, and ^{60}Co. Such isotopes are a powerful diagnostic of the explosion, telling us about conditions within the star, the extent to which convection mixes the exploding star, and so forth. Gamma-ray lines from the ^{44}Ti decay chain were recently observed. To interpret such measurements, one must understand how such elements are synthesized in the explosion, how they decay, and how that synthesis depends on variables such as the mass of the progenitor. This is a rich area for collaboration between nuclear physicists and astrophysicists.

Finally, one of humanity's most exciting endeavors in the twentieth century has been its ventures into space. The cosmic-ray environment of space is more severe than on Earth, and the associated radiation hazards to human health must be evaluated and addressed. Nuclear science is contributing to this effort by studying radiation damage under the controlled conditions of accelerator beams, providing the basic knowledge needed for the interpretation of space-bound measurements.

NEUTRON STARS

Neutron stars are the extraordinarily dense stellar cores that remain after supernova explosions. They can be studied through their radio, optical, x-ray, and gamma-ray pulsations, as well as through their gravitational effects on companion stars. They typically contain about 1.5 times the mass of the Sun, packed into a sphere only 10 kilometers in radius. The enormous gravitational forces in neutron stars compress electrically neutral matter to densities up to approximately 10^{15} g/cc, or about five times that of charged nuclear matter in atomic

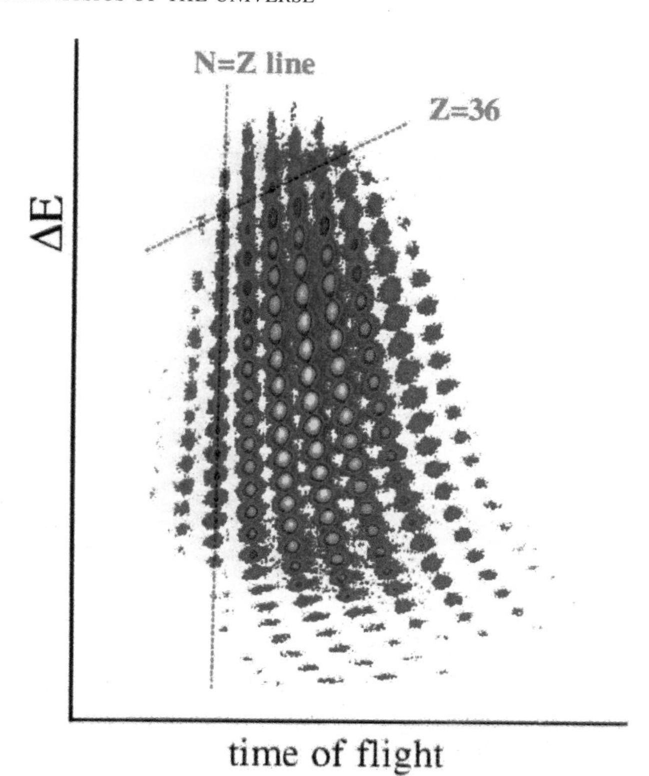

FIGURE 5.6 The *rp* process can be halted by encountering a nucleus that too easily emits protons, and thus lives only a very short time. Nuclides resulting from the breakup of ^{78}Kr in an in-flight radioactive beams facility are shown in the panel above. They were detected and identified from their velocity and charge. The proton-rich nuclei that participate in the *rp* process lie to the left of the $N = Z$ line. Their appearance on the plot indicates that they live sufficiently long to be of importance to the *rp* process.

nuclei. Thus, a teaspoon of neutron-star matter would have the same mass as a cubic kilometer of matter on Earth. There is a maximum mass for neutron stars; compact objects exceeding this maximum value collapse under their own gravity into black holes.

Nuclear theorists have been able to exploit neutron stars as an important laboratory for the study of dense matter. For example, the limiting mass of neutron stars, beyond which black holes are formed, is enhanced by nuclear forces, which resist gravity, from 0.7 times to over twice the mass of the Sun. Thus, nuclear forces are essential for the existence of the commonly observed neutron stars having 1.4 times the solar mass.

Solid metallic iron, with the lowest energy per nucleon of all forms of matter, constitutes the outer surface of neutron stars. But as one descends below the surface, the matter is squeezed more and more tightly by gravity (Figure 5.7). Initially, this does not affect the nuclear properties of the matter; the compression works on the electron gas surrounding the nuclei. But beyond a density of about 8×10^6 g/cc, the electron energy becomes so great that electrons begin to be captured by protons in the nuclei, transmuting the iron into nuclei that are more neutron-rich. This process continues with further compression until matter reaches a density of about 4×10^{11} g/cm^3. Up to this point, the matter has the familiar form of a crystal lattice of nuclei surrounded by a dense electron gas.

Beyond this density, the additional neutrons produced by electron capture do not remain bound to nuclei; like electrons, they form a gas that fills the space between the nuclei. This new form of matter is still a solid, but it may have novel

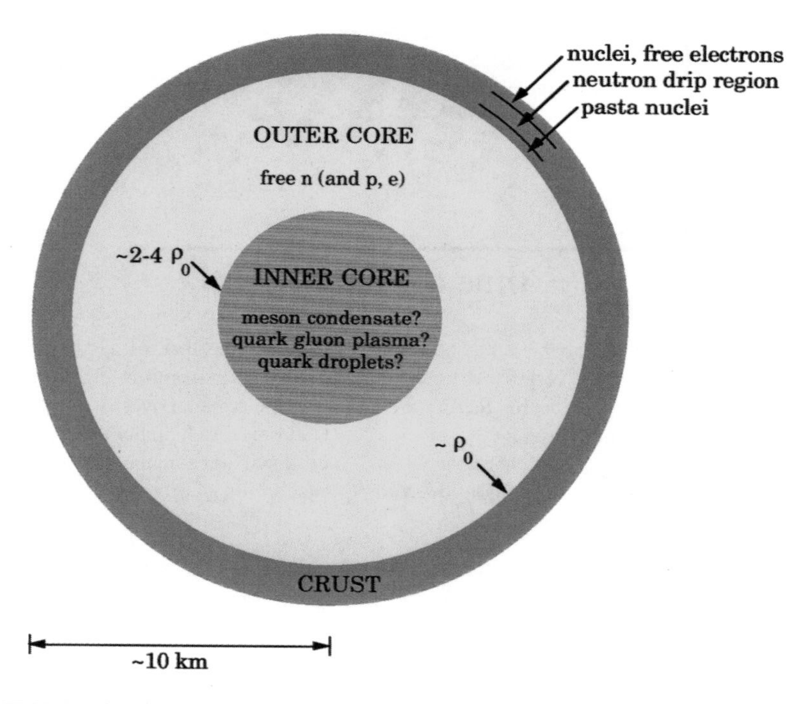

FIGURE 5.7 Schematic cross section of a neutron star, illustrating possible states of the hadronic matter as one progresses from the crust to the inner core, where densities perhaps four times that of ordinary nuclear matter will be encountered. The crust is solid metallic iron, while the outer core is a neutron gas, with a small admixture of protons and electrons. At the extreme densities characterizing the inner core, many exotic forms of nuclear matter—pion or kaon condensates, quark droplets, or a quark-gluon plasma—could exist.

features. For example, when sufficiently compressed, the nuclei of this matter could join to form thin and long spaghetti nuclei, which in turn might merge at even higher densities into thin flat sheets of "lasagne." Finally, at a density of about 10^{14} g/cc, the nuclei are fully dissolved and matter becomes a nuclear fluid.

The evolution of the matter traced above—from solid iron to fluid—occurs as one descends through the crust of a neutron star, a layer about one kilometer in thickness. The neutron gas in this crust is believed to be superfluid, and the puzzling phenomena of glitches, i.e., sudden speedups observed in the rotational periods of neutron stars, are likely due to the transfer of angular momentum from the superfluid to the remainder of the crust.

The fluid matter at densities of the order of 10^{14} g/cc contains mostly neutrons, with a small admixture of protons and electrons. Deeper within the neutron star, where gravity compresses the matter to densities approaching 10^{15} g/cc, more exotic changes can occur. Although accurate predictions are difficult to make, given our uncertain knowledge of nuclear forces in this regime, the possibilities include:

- Some of the most energetic electrons in matter are likely converted to muons, heavy cousins of electrons in the family of charged leptons. It is also possible that the electrons are replaced in part by negatively charged pions or kaons, condensed into quantum states exhibiting interesting collective properties.
- Some of the most energetic neutrons or protons may be converted into heavier baryons, such as lambdas or sigmas, containing strange quarks.
- The neutrons and protons may dissolve into uniform quark matter or quark-matter droplets, the low-temperature form of the quark-gluon plasma discussed in the last chapter.

Two important aspects of this dense fluid are its equation of state and composition. The equation of state determines the relationship between the mass and the radius of a neutron star. Masses of a number of neutron stars are known from their orbital motion about other stars, and indirect means exist for deducing radii from observations. The composition is important to neutron-star cooling: hot, nascent neutron stars, produced in supernova explosions, cool through neutrino emission processes that take place in the matter. Measurements of the surface temperatures of neutron stars may be possible with satellite-based x-ray observatories within the next decade. Such measurements could tell us whether exotic components of nuclear matter—constituents other than neutrons, protons, electrons, and muons—are playing a role in the cooling. Finally, one of the major goals of the Laser Interferometer Gravitational-wave Observatory (LIGO), now under construction, is to measure gravity waves emitted in the coalescence of two neutron stars. The pattern of the waves emitted just prior to merger is also sensitive to the structure of the stars and the equation of state.

Efforts to understand dense nuclear matter and to predict the properties of neutron stars depend on knowledge of nuclear interactions gained in the laboratory. In the next few years, new progress is expected. Heavy-ion collisions will help us better understand the interactions of mesons in hot, dense nuclear matter, which is crucial to the issue of meson condensation in neutron stars. Future studies of neutron-rich nuclei, near the limit of stability, in radioactive ion beam facilities will allow us to more accurately model nuclear forces in neutron star crusts.

PARTICLE PROPERTIES FROM NUCLEAR ASTROPHYSICS

Some of the nuclear astrophysics issues discussed above have important implications for the properties of elementary particles. The Big Bang limit on the number of light neutrino species and the implications of the solar neutrino puzzle for neutrino mixing are two famous examples. Another important set of particle physics constraints comes from studying how stars cool. Often there exists a delicate balance between some temperature-dependent nuclear reaction rate and the life cycles of stars. For example, red giant core ignition, discussed above in connection with the fusion of three ^{4}He nuclei into ^{12}C, can serve as an extremely sensitive stellar thermometer, because a 1 percent change in core temperature leads to a 30 percent change in the triple ^{4}He reaction rate; this is known from laboratory measurements that constrain the nuclear cross sections. Thus, the ignition of ^{4}He can be appreciably delayed by even modest stellar cooling associated with, for example, anomalous electromagnetic interactions that produce neutrinos. Such a delay has consequences for astronomy, altering the ratio of red giant to horizontal branch stars (the stage of slow stellar evolution following the red giant phase). This chain of reasoning provides our best experimental constraint on magnetic moments of neutrinos, a limit that is a factor of 100 more sensitive than those derived from direct laboratory measurements.

Similarly, one type of neutrino mass can allow neutrinos to scatter into so-called sterile states in which no interactions with matter can occur. Neutrinos could then immediately escape from a supernova. If the cooling is increased too much, a conflict with the measured duration of the SN1987A neutrino pulse arises. Nuclear physics is crucial in deriving such limits, which are extremely sensitive to the maximum temperature achieved after core bounce. Thus, it is essential to understand the uncertainties in the nuclear equation of state when extrapolated to the unusual temperatures and densities encountered in the cores of supernovae.

Similar arguments apply to new particles that might mediate the cooling. One important example is the axion, a particle predicted in theories that explain why the strong interaction respects time-reversal symmetry so exactly. The combination of red giant and supernova cooling arguments limits possible axion masses and couplings to a narrow range, a window where such axions might also be an important contributor to dark matter.

OUTLOOK

Present efforts in nuclear astrophysics may soon lead to the solution of the solar neutrino problem, the successful modeling of the supernova explosion mechanism, an understanding of the nucleosynthesis of heavy elements, and more quantitative constraints on the structure and dynamics of neutron stars. If the solution of the solar neutrino problem involves massive neutrinos, nuclear physics will have demonstrated the need for physics beyond the current Standard Model of particle physics. The supernova mechanism and the origin of the heavy elements are questions with deep connections, as the dynamics of the supernova explosion and the fossil record of that explosion in the synthesized nuclei must be understood within a single model.

In the longer term, it is apparent that an explosion of new instrumentation is revolutionizing the rich intersections between astronomy, astrophysics, and nuclear physics: new technology telescopes, satellite-based detectors for probing the microwave background and for measuring astrophysical sources of gamma rays, more sophisticated solar-neutrino detectors, radioactive-beam facilities for the study of reactions that previously occurred only in stars, high-energy neutrino detectors utilizing the ice caps or the oceans, new observations of neutron stars (including the exciting prospect that the detection of gravitational radiation emitted when neutron stars collide may become possible), and others. Increasingly, the understanding of new data on astrophysical objects depends on our understanding of the underlying nuclear (and atomic) microphysics that drive the evolution and energy production of the stars and the galaxies in which they reside. This inextricable linking of nuclear physics and astrophysics has produced a rich bond between the two fields that seems destined to grow ever stronger.

6

Symmetry Tests in Nuclear Physics

INTRODUCTION: PRIORITIES AND CHALLENGES

The realization that symmetries are the key to understanding and classifying the structure of matter and the fundamental forces is among the deepest theoretical insights of this century. The seminal work is identified with the Nobel Prize-winning nuclear theorist, Eugene Wigner, whose ideas have influenced every area of modern physics. The structure and behavior of nuclei reflect the symmetries of the basic building blocks and their interactions. Nuclei can also be laboratories for studying the symmetries of the fundamental processes themselves.

High-energy particle physics grew out of early nuclear physics, so it is not surprising that there are similarities between some of the scientific goals of the two fields. Each field brings its own arsenal of experimental and theoretical techniques. Understanding the fundamental forces, the main focus of particle physics, is an essential step toward a deep understanding of the nucleus. But the nucleus can be a powerful tool for studying fundamental forces themselves. To exploit the nucleus as a research tool, experimentalists capitalize on the understanding of the structure of specific nuclei accumulated over the last sixty years. Successful nuclear models are the instruction manuals for understanding this nuclear laboratory. There has been enormous progress in the last two decades, and amazing experimental feats have been accomplished: nuclear physicists have measured neutrino scattering from specific nuclear states, observed the incredibly weak process of double beta decay, and detected neutrinos coming from the Sun despite the feeble strength of the basic interactions of neutrinos at low energies. There are a number of challenging questions for nuclear physicists at the particle physics boundary:

- Do neutrinos have mass? Are the leptons from different generations mixed together by the weak interaction?
- Is the pattern of symmetry violations found experimentally at low energies consistent with the Standard Model description? Can a direct manifestation of a violation of time-reversal symmetry be discovered at low energies?
- What is the significance of the mixing of quarks from different generations? Will more precise experiments show that mixing parameters satisfies the quantum mechanical laws of probability within the context of three generations of quarks and leptons?

THE STANDARD MODEL

The first hint of the existence of the fundamental weak force was the discovery of radioactivity about a century ago. It has taken most of the last hundred years to establish the Standard Model, a surprisingly complete description of the fundamental particles and their interactions. The most important clues came from elegant experiments that studied weak-interaction phenomena. The Standard Model describes the electromagnetic, weak, and strong interactions; gravity is the only known interaction missing. While there are occasional hints of the need for revision, no confirmed experiment contradicts the Standard Model. Nevertheless, despite the successes, most physicists are convinced that a more complete theory of nature will eventually replace it. The Standard Model has a disturbingly large number of parameters whose numerical values are not explained; many aspects of the model seem unnatural.

Since the early 1970s, experimentalists have mounted a two-pronged assault on the Standard Model attempting to discover its limitations: trying to verify its quantitative predictions to the highest possible precision, and searching for new, unexpected, and inconsistent phenomena. The top priority of high-energy physics is to discover direct evidence for the Higgs boson, the particle responsible for all finite particle masses according to the Standard Model. If there is no Higgs boson, then the Standard Model must be revised. The prudent search for the Higgs is at the highest available energy. Searches for specific extensions to the Standard Model (for example, a class of attractive theories classified as grand unified supersymmetric models) are often, but not always, carried out at high energies. Nevertheless, it may turn out that testing the Standard Model precisely at low energies is the most economical way to find new physics. Low-energy experiments often have unique discovery potential.

The best-established part of the Standard Model is its description of the electromagnetic interaction. Called quantum electrodynamics (QED), it developed well before the weak and strong interaction were incorporated into the modern Standard Model. QED is the theoretical prototype of the entire Standard Model. Confidence in QED comes mostly from atomic physics tests. Although no longer a major thrust, precision tests of QED continue with nuclear physicists participating in important areas. One aspect of the most interesting modern work involves a simple but exotic atom, muonium, consisting of an antimuon (the antiparticle of the muon) and a bound electron. Elegant muonium experiments have been done by nuclear and atomic physicists who exploit medium-energy nuclear physics accelerators, which are muon sources. Nuclear physicists are working with particle and atomic physicists on major new experiments to determine precisely the magnetic moment of the muon (to be precise, the quantity $g - 2$). The muon's $g - 2$ is a measure of the QED corrections to the quantitative relationship between the muon's spin and its magnetic moment. In principle, theory can predict $g - 2$ exactly, but calculating the value to better and better precision is a formidable theoretical task. For decades, newly developed experimental and theoretical techniques have challenged each other, searching for a discrepancy that might indicate that the underlying theory is wrong. Modern experiments rely on sophisticated accelerator technologies, and muon, which have a lifetime of only two-millionths of a second, are confined in specially constructed storage rings. These experiments have the unique capability of testing particular Standard Model predictions. But the main thrust of the nuclear physics assault on the Standard Model focuses on the fundamental weak interaction. An active area of research involves detailed studies of the phenomena of nuclear beta decay, the process responsible for natural radioactivity.

The elementary particles of the Standard Model are summarized in Figure 6.1. The fundamental constituents of matter are the structureless quarks and leptons and the forces between them that arise from the "exchange" of particles called gauge bosons. Photons, gluons, and the W and Z bosons are the gauge bosons for the electromagnetic, strong, and weak forces, respectively. The interactions are characterized by special kinds of symmetry classified as gauge symmetries. In a quantum theory, a symmetry is a specific change of variables that leaves all the physical predictions unchanged. Each symmetry implies an associated conservation law obeyed by the interaction. The correct theory of nature must account for the conservation laws that are verified experimentally. For example, one symmetry of electrodynamics implies that the photon has no mass and that electric charge is conserved. No experiment has ever found a violation of electric-charge conservation or a nonzero photon mass, and this symmetry is well established. The fact that photons themselves do not have electric charge makes QED rather simple. The strong force described by quantum chromodynamics (QCD) is mediated by massless gluons, but they are self-interacting; consequently, QCD is much more complicated than QED. One important implication of self-interacting gluons is that under normal conditions, quarks are confined. The basic building blocks of all hadrons and ultimately the atomic nucleus, the quarks, apparently cannot exist in isolation. In particular, no experiment has

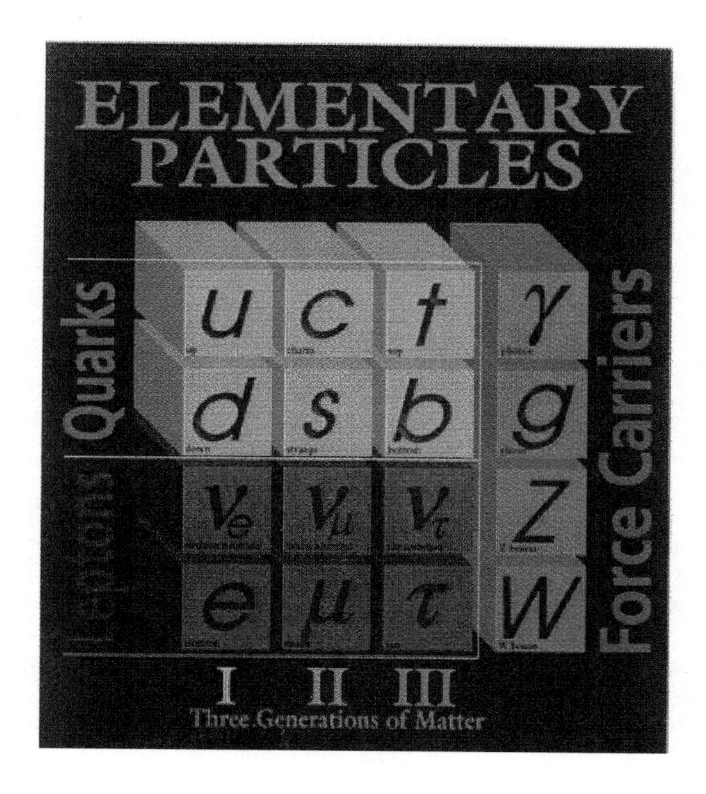

FIGURE 6.1 The few basic building blocks (quarks and leptons) and the exchange particles (gauge bosons) of the Standard Model are economically displayed in a chart much smaller than the periodic table of the elements.

succeeded in isolating a quark. While it is believed that quark confinement is fully consistent with the Standard Model, a rigorous theoretical demonstration has not been made because of the complications of dealing with a nonlinear theory.

Beta decay is a process in which a nucleus emits an electron and a neutrino and results from the exchange of a particle, in this case a massive *W* boson. In beta decay, one of the quarks transforms into a quark of a different type (a different flavor) and at the same time a *W* boson is emitted, which quickly decays into an electron and a neutrino (electrons, muons, tauons, and neutrinos are the elementary particles that are classified as leptons). The character of the particular interaction is determined by the properties of the exchanged particles; the range of the interaction depends on their mass. The weak interaction is a weak force with a short range because the *W* and *Z* gauge bosons are rather heavy, about 90 times heavier than the proton.

The Standard Model has already stood up against decades of intense experimental scrutiny. Why, then, is it not regarded as the best possible theory of nature, "the ultimate theory"? In spite of its successes, the Standard Model has about 19 parameters that are apparently arbitrary—numbers whose origins are not understood. Moreover, some of the parameters have values that are hard to understand. For example, in the present Standard Model, neutrino masses are all set to zero (in fact, the model would not require a big change to incorporate finite neutrino mass). The situation is much different from that of the massless photon, because there is no obvious symmetry in the Standard Model that accounts for massless neutrinos. In general, the particle masses are modified by interactions. Keeping physical neutrino masses zero seems to require a conspiracy: the magnitude of the bare neutrino masses and the strength of the interactions must be almost exactly tuned to give zero in the end. Physicists have come to regard theories that must be finely tuned as unnatural and unsatisfactory.

All the quarks are assigned a one-third unit of baryon number. So far, baryon number is absolutely conserved according to experiment. But, as with the zero neutrino masses, no underlying symmetry explains baryon-number conservation. If there were a symmetry, baryon number should be associated with a force, like electric charge, but experiments have found no force that couples to baryon numbers. A similar problem occurs for leptons: experiments seem to require conserved lepton numbers (lepton number discriminates between leptons from different generations) in the absence of an underlying symmetry. With no good reasons for these conservation laws, searches for violations of baryon and particularly lepton number continue to be actively pursued by nuclear physicists.

The unsatisfactory features of the Standard Model suggest specific areas for experiments to probe. For example, looking for conclusive evidence of nonzero neutrino mass is a major goal of nuclear physics.

TESTING SYMMETRIES

Understanding the symmetries relevant to the behavior of the physical world, even approximate symmetries, organizes the underlying physics. The connection between symmetries and conservation laws is the key link in this understanding. Conservation laws for energy and momentum are understood as a consequence of underlying symmetries with respect to translations in time and space. If the laws of nature are the same everywhere in the universe and do not change in time, then energy and momentum must be conserved absolutely. The conservation laws for energy and momentum are so well tested and so well established that it is almost inconceivable that a violation will ever show up.

As experimental and theoretical tools grew more sophisticated, symmetries other than those of space-time were introduced in physics. These go under the name of internal symmetries. Electric-charge conservation is one example. Some related internal symmetries first studied in nuclear physics include charge independence and charge symmetry. Neutrons, being neutral particles, are not affected much by the electromagnetic force. On the other hand, protons, being charged, feel a much larger electrical force. Charge symmetry is a statement about the strong force: the nuclear force between two neutrons is the same as the nuclear force between two protons. Charge independence is a bigger symmetry, relating a proton-neutron system to a proton-proton or neutron-neutron system: the nuclear force between any two nucleons (be they protons, neutrons, or a combination) is the same, apart from the electroweak contributions. The ideas of charge symmetry and charge independence in nuclear physics led to the marvelous concept of isospin. Isospin is an abstract way of describing the symmetry between neutrons and protons. The theoretical description of isospin is mathematically similar to angular momentum and spin, but isospin is in a completely abstract space in which the axes have something to do with the type of particle or the flavor. This revolutionary theoretical idea is the prototype of techniques for describing all internal symmetries. Charge symmetry and charge independence are not exact symmetries in nuclear systems, and accounting for the observed deviations tests our understanding of the limits of this symmetry. Isospin symmetry is broken by the electromagnetic interaction and because each type of quark has a definite mass. Understanding broken symmetries better is an ongoing goal of modern nuclear physics; the particulars of symmetry breaking are studied with careful measurements of polarization effects in reactions with light nuclear systems at medium-energy nuclear physics accelerators. One recent experiment (shown in Figure 6.2) employed sophisticated methods of polarizing liquid hydrogen targets and highly polarized proton beams.

For a long time, it was assumed that the laws of nature are unchanged by mirror reflection: a process seen in a mirror should be completely consistent with its theoretical description. Experiments had already established that other spatial transformations—translations and rotations—are indeed symmetries obeyed by the laws of physics; furthermore, no evidence for inversion symmetry breaking was evident in the structure of atoms or nuclei, so it was assumed that all physical laws obey mirror symmetry. It came as a tremendous surprise in 1956 when it was discovered that the weak interaction violates mirror symmetry. The associated quantum number, parity, is not conserved in certain fundamental processes. The first clear experimental demonstration was done by a group of nuclear physicists. They studied the beta decay of ^{60}Co that had been diffused into a paramagnetic salt and spin-polarized in a high magnetic field at low temperatures. It was found that electrons emitted in this decay are emitted preferentially in the direction of the nuclear spin. But when viewed in a mirror, the relative orientation of the electron momentum to the nuclear spin reverses, as shown schematically in Figure 6.3. The observed asymmetry in the ^{60}Co beta decay means that we would be able to tell whether we were viewing the process directly or in a mirror! The observation of parity violation caused physicists to question whether or not other discrete transformations—conjugation of charges (exchanging particles for anti-

FIGURE 6.2 The apparatus used at the Indiana University Cyclotron Facility (IUCF) to measure small violations of charge symmetry in nuclear forces. The experiment was a high-precision study of the scattering of a spin-polarized neutron beam by a spin-polarized proton target. The scattered neutrons and protons were detected in the counters shown. (Courtesy Indiana University Cyclotron Facility.)

particles) and reversal of time (the process evolving in reverse)—are symmetries of nature. It was quickly realized that charge-conjugation symmetry is violated as badly as reflection symmetry.

Spatial Reflection Symmetry

Experiments suggest that reflection symmetry is broken as badly as it could be in processes like beta decay. This observation is incorporated into the Standard Model, but what is the reason for maximal parity violation? There are alternative models to the Standard Model—for example, attractive left-right symmetric models that contain an extra set of intermediate-vector bosons with the other sense of parity violation. Consistency with low-energy experiments is obtained if the extra bosons have masses sufficiently large not to have an effect. Nuclear physicists are instrumental in testing this plausible class of Standard Model extensions. Precise measurements of basic parity-violating effects in the

FIGURE 6.3 Schematic diagram of the principle of the first experiment to discover that reflection symmetry is violated by the weak interaction. A ^{60}Co nucleus has many of the properties of a spinning top. If the electron from the beta decay of ^{60}Co were to be emitted along one sense of the axis of the spinning nucleus, then the situation would be just the opposite in a mirror. Nuclear physics experiments have clearly demonstrated that reflection symmetry is not obeyed by the weak interaction.

ordinary beta decay of muons and nuclei are compared to the exact Standard Model predictions. Nuclear physicists have invented novel techniques for polarizing nuclei and measuring the positron helicities to accomplish these tests. The neutron is also an excellent system for study, because it is possible to achieve a high degree of neutron polarization and the neutron is simple and more easily understood theoretically than are more complex nuclei. Presently, the most precise experiments exploit muon beta decay and ordinary nuclear beta decay. The muon experiments have significant prospects for improvement because of new initiatives for intense low-energy muon sources at meson factories. For nuclear beta decay, the rates are usually high; improving the sensitivity is a matter of eliminating systematic uncertainties. The latest experiments attempt to study the beta decay of radioactive atoms in the extremely clean environments of optical traps. Optical trapping is a way of confining atoms to a small region of space by using the radiation pressure of laser light. These experiments promise enormous improvements, but the level of precision of existing experiments is good enough to rule out theories with extra exchange gauge particles as heavy as several hundred GeV compared to the 90-GeV *W* and *Z* particles of the Standard Model. Currently, the limits from low-energy experiments are often beyond those from direct searches at high-energy accelerators.

Time-Reversal Symmetry

Reversal of time is the only discrete transformation of space-time that has not been demonstrated to be broken. A violation of time-reversal symmetry means that nature favors the course of a fundamental process to proceed with a particular direction of time. If nature were symmetric under time reversal, we could not determine if a movie of a fundamental process were running forward or backward. It is natural to have a sense of the direction of time. For example, people age, beaches erode, and crops grow; the reverse is unnatural. These kinds of time-symmetry violation are understood as the natural tendency of complicated systems to evolve toward a more and more disordered state, but at the level of the fundamental interactions, the significance of an arrow of time is much more important. A deeper understanding of time-reversal invariance may hold the key to understanding the origin of the universe. It is convincingly argued that time-symmetry breaking is an essential ingredient for creating a universe that is primarily matter instead of equal mixtures of matter and antimatter. Nuclear physicists probe this question with experiments searching for the electric-dipole moments (EDMs) of neutrons and atoms. Static EDMs are forbidden in the absence of time-symmetry violation. In the case of atoms, an EDM can arise if the electron has an electric-dipole moment or if there is a time-symmetry-violating interaction within the nucleus. The most recent experiment to search for the electric-dipole moment of the neutron is shown in Figure 6.4. The limits on the existence of atomic and neutron EDMs provides some of the best constraints on some of the most plausible extensions to the Standard Model. One class of possible extension is called supersymmetric theories, which predict the existence of a large number of new and as yet undiscovered particles. Important constraint on many of these new theories comes from searches for Standard Model-forbidden rare decays of muons. The most precise limits on rare muon decay comes from experiments carried out at the Los Alamos Meson Physics Facility (LAMPF) during the last two decades. Other constraints come from low-energy experiments searching for EDMs of electrons and nucleons.

It is also possible that time-reversal-symmetry breaking might show up in the decay of an unstable system. Modern experiments search for T-violating correlations in the beta decay of neutrons, mesons, and particular nuclei. A new search for a T-violation effect in neutron beta decay is now under way at the National Institute of Standards and Technology (NIST) reactor in Gaithersburg, Maryland, shown in Figure 6.5. Another precise limit on T-violation comes from the beta decay of ^{19}Ne. The experiment uses the atomic beam technique applied to radioactive atoms produced online at a cyclotron. A related transformation,

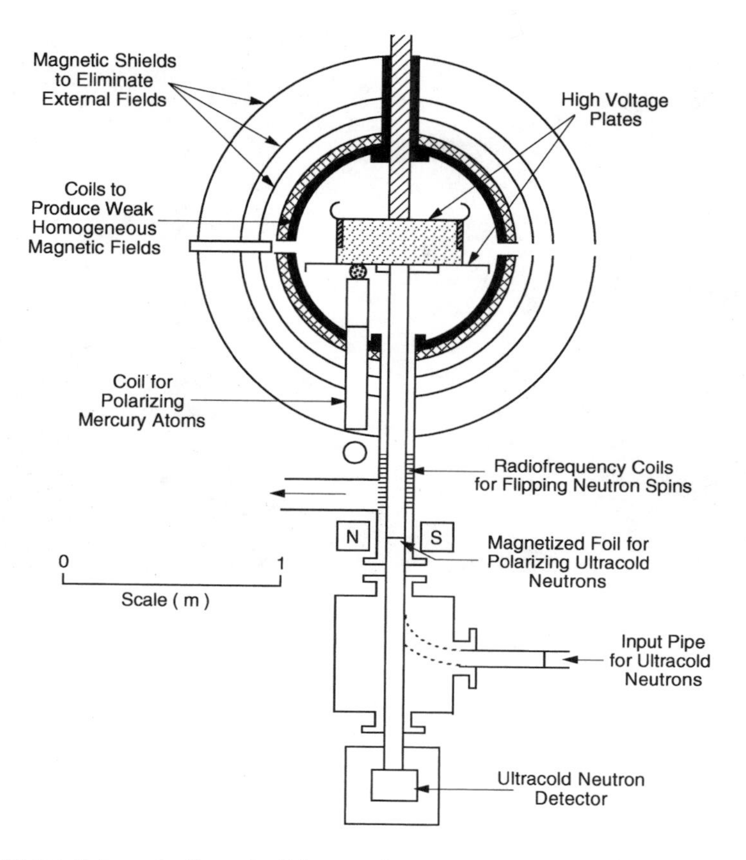

FIGURE 6.4 Schematic diagram of the experiment to search for a neutron electric-dipole moment at the Laue-Langevin Institute (ILL) in Grenoble, France. The experiment exploits the unusual properties of ultracold neutrons. Ultracold neutrons can be confined in material bottles and polarized with thin magnetized foils. This experiments searches for a change in the neutrons' precession frequency caused by an imposed electric field. The neutrons' precession is directly compared to the precession of an isotope of mercury, which should show no effect. The effect of an electric field on the neutrons' precession would indicate an electric-dipole moment and a violation of time-reversal symmetry. (Courtesy of Laue-Langevin Institute.)

called CP, is obtained by combining mirror reflection with the exchange of particles with antiparticles; CP was discovered to be a broken symmetry at high-energy accelerators during the 1960s. The effect is small, showing up in particular decay modes of the neutral K-mesons. Theory makes a strong case that CP violation implies T-violation, but so far no explicit T-violation has been observed. The answer to this puzzle will be a major advance in the quest for a better

FIGURE 6.5 An experiment to study how well time-reversal symmetry is obeyed in the beta decay of the free neutron. A beam of cold neutrons enters the detector region form the upper right. The detector is situated in a carefully designed magnetic guide field. Some of the detector elements for measuring the decay electrons and residual protons from neutron beta decay are visible to just downstream of the experimentalists shown in the photograph.

theoretical description of nature. The Standard Model incorporates T-violation in a set of parameters that are taken from kaon-decay experiments, but the size of any direct T-violation effects is predicted to be too small to measure with present experimental techniques. If T-violation were found at the level of sensitivity of present experiments, it would already be evidence for a breakdown in the Standard Model.

PRECISION MEASUREMENTS OF STANDARD MODEL PARAMETERS

Precisely measuring the Standard Model parameters is essential, as one of the best ways of establishing the consistency (or inconsistency) of the theory. All experiments, independent of the energy of the process involved, must yield exactly the same value of the particular parameter. Nuclear physics experiments play an important role in this program by establishing consistency at low energies.

For reasons still unknown, the weak interaction mixes quarks together (in the sense of quantum mechanics) in peculiar ways. Four Standard Model parameters characterize this mixing. One parameter is measured in ordinary nuclear beta decay and others in the beta decays of particles containing strange and bottom quarks. The determination of the weak vector coupling constant from nuclear beta decay is crucial. The experiments involve measurements of masses, decay energies, and branching ratios, among the most precise measurements in nuclear physics. This work exploits the most sophisticated nuclear physics devices available; for example, a recent experiment exploited the Gammasphere, one of the world's best tools for elucidating the nuclear structure of nuclei with large angular momentum. Small corrections from the effect of the electromagnetic interactions within the nucleus must be made with extremely good precision, using the most sophisticated shell-model representations of the nucleus. At the level of present experiments a tantalizing hint of a discrepancy with the Standard Model is emerging, but better precision is required before a firm conclusion is possible. Studies of neutron beta decay can also shed light on this issue, but the precision is not quite up to the level of the experiments using nuclear decays. These same neutron-decay experiments are presently the best way to measure the nucleon axial vector coupling, which should eventually be predicted with precision by QCD. Neutron-decay experiments are expected to improve with the availability of more intense sources of cold and ultracold neutrons being planned at future neutron-spallation facilities. Meanwhile, particle and nuclear physicists have collaborated to mount an experiment to improve the precision of the mixing parameter that comes from strangeness-changing beta decay by measuring the lifetime for a kaon to beta decay into a pion. In Europe, a U.S.-led group is studying the beta decay of the pion at a large nuclear physics cyclotron in Switzerland. From the theoretical standpoint, pion decay is perhaps the cleanest way to study this particular Standard Model parameter, but the experiment is extraordinarily difficult because of the small probability that a pion decays in the required way.

Beta-decay experiments can be designed to search for unexpected forces, and it is likely that the old process of nuclear beta decay will remain a valuable tool for a long time. Techniques for storing neutrons, atoms, and ions in traps hold the promise of experimental breakthroughs. New intense sources of cold and ultracold neutrons and intense sources of exotic radioactive nuclei at future exotic-beam facilities will make this an even more active research area in the coming years. Figure 6.6 shows about 50,000 radioactive ^{21}Na atoms being used to study reflection symmetry. This experiment should help answer the puzzle of whether or not parity is violated as badly as it could be in processes like beta decay. This technique will have even more applications at future exotic-beam facilities capable of producing even larger quantities of short-lived radioactive atoms.

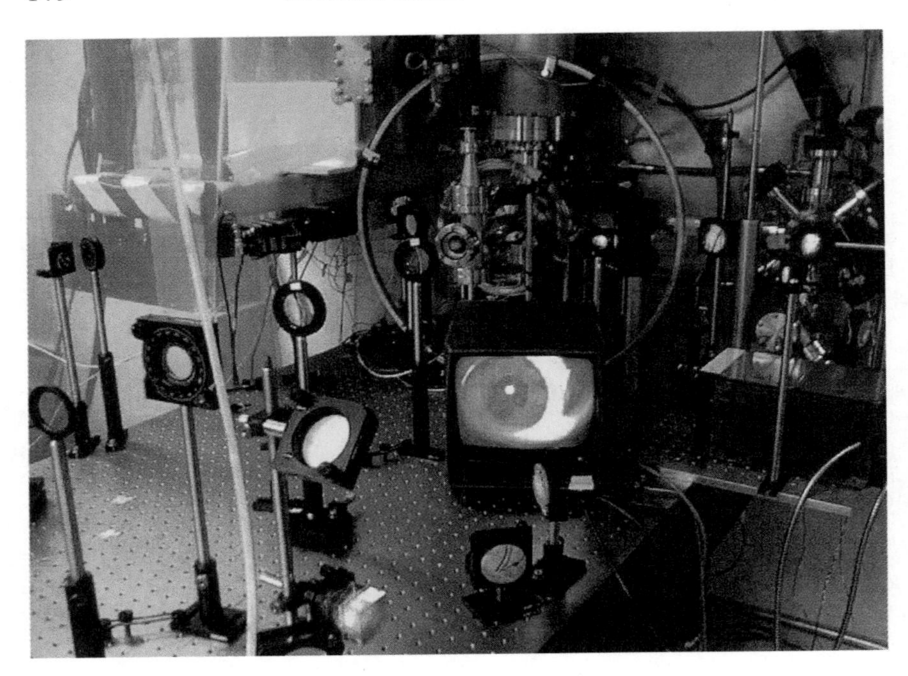

FIGURE 6.6 The monitor in this photograph is displaying about 50,000 atoms of radioactive ^{21}Na that are confined in a specially designed atom trap on the right. The isotope ^{21}Na has a half-life of only 22 seconds so the atoms must be continuously produced with a cyclotron accelerator. Collecting radioactive atoms and studying their decays is an excellent method of testing the symmetries implied by the Standard Model.

One of the best ways to measure the mixing between the weak and electromagnetic interactions, the Weinberg angle, is to study the parity-violating interaction between electrons and the nuclei of particular atoms. Parity mixing has been seen in several atomic systems; the best measurement at present is done with ^{133}Cs atoms. Nuclear physicists are exploring the possibility of measuring this effect in atomic francium, where the parity-mixing effect should be about 18 times larger. Since there are no stable isotopes of francium, the experiment must be carried out with a small number of radioactive atoms. Recently, nuclear physicists have managed to collect francium in a magneto-optic trap. Studying parity mixing in a series of isotopes of the same atom is an attractive method of making the determination of the Weinberg angle at low energy more reliable. One way to carry out such a program is to do these experiments with radioactive atoms.

THE SEARCH FOR NEUTRINO MASS

Nuclear physics played a key role in establishing the existence of the mysterious neutrino that was originally proposed to explain the apparent lack of energy conservation in nuclear beta decay. Unique among all the elementary constituents of the Standard Model, the neutrino interacts exclusively by the weak interaction, making its direct detection extremely difficult. It was first observed directly by experimentalists who used a nuclear reactor as a neutrino source. From the beginning, the neutrino was expected to be very light. Enrico Fermi was first to point out that neutrino mass could be determined from systematic studies of electrons from nuclear beta decay. The masses of neutrinos are taken to be zero in the Standard Model but, as was noted above, there is no natural explanation for zero-mass neutrinos. Finding out whether or not neutrinos have mass is among the most important issues in modern physics. Finite neutrino mass would require a modification of the current Standard Model, and present a number of intriguing cosmological implications.

Nuclear physicists have refined the method of measuring neutrino mass from the beta decay of tritium, Fermi's original suggestion. The conclusion of a half-dozen experiments in the last decade has been that the electron neutrino has a mass no larger than 10 eV, or less than 1/50,000 of the electron's mass. These experiments have reached a level of accuracy at which subtleties of atomic and molecular physics are important systematic uncertainties. A better understanding of the molecular and atomic physics involved is needed before the existing data can be used to put more stringent limits on the neutrino mass. Figure 6.7 is a photograph of one of the large spectrometers built to study the mass of the neutrino on the basis of tritium beta decay. At present, a half-dozen direct searches for neutrino mass from the decay of tritium have found no evidence for finite neutrino mass. However, the high-precision data are not completely understood, and all the experiments indicate a systematic deviation from the normal theory of beta decay. The origin of the deviation may have a simple explanation; for example, an unappreciated complication of the molecular physics of the radioactive tritium source, or it may be a clue that will lead to new physics. Efforts to resolve this issue are continuing.

A fascinating process that would cause neutrinos of one type to change into another is at the heart of what is evolving into a major scientific advance. The possibility that neutrinos of different types might transform into one another is allowed by the quantum nature of neutrinos. Such neutrino oscillations would occur if at least one of the neutrino types had mass and if neutrinos were to mix in a way similar to the quarks. Thus, neutrino oscillations offer another way to look for finite neutrino mass. In the absence of definite predictions about the likely masses and mixings of neutrinos, and in view of the solar neutrino results already discussed above, a number of experiments with different sensitivities are going on at high-energy accelerators, lower-energy nuclear-physics accelerators,

FIGURE 6.7 A large beta spectrometer (7 feet in diameter) designed especially to measure the energies of electrons emitted from the beta decay of radioactive tritium, an isotope of hydrogen with two neutrons and one proton. The photograph shows the enormous magnetic field coils in this toroidal spectrometer. By carefully studying the spectrum of electrons from tritium decay, it is possible to detect whether or not the neutrinos, also emitted in the beta-decay process, have a mass that is different from zero. (Courtesy of Lawrence Livermore National Laboratory.)

and reactors. Other important experiments use natural sources of neutrinos either from the Sun or from the collisions of cosmic-ray particles in the upper atmosphere. There are exciting results from a number of these experiments. The evidence of neutrino mass from atmospheric neutrinos and solar neutrinos seems particularly convincing, but there are also positive indications from an accelerator experiment.

An experiment using neutrinos produced in the beam stop of the LAMPF proton linac seems to show evidence of muon antineutrinos oscillating into electron antineutrinos, but other experiments have not reproduced this evidence yet. A photograph of the inside of the 100-ton liquid scintillator detector used in this experiment at LAMPF is shown in Figure 6.8. The neutrino mass hinted at, if this result were correct, is rather large and tantalizing because there could be cosmological and astrophysical implications. A sufficiently large neutrino mass might help explain the puzzling astrophysical evidence that much of the matter in the universe is not ordinary hadronic matter. Fortunately, a similar oscillation experiment in England may provide confirming evidence in the next few years. Experiments proposed at higher-energy accelerators should be capable of seeing the same effect with even better sensitivity. The other indications of neutrino mass using natural neutrino sources suggest smaller mass scales; while this does not directly contradict the LAMPF result, a fairly drastic modification of the Standard Model is necessary in order to accommodate all the positive indications.

The earliest neutrino oscillation experiments were carried out by nuclear physicists at reactors. With reactor sources one searches for the disappearance of electron neutrinos into any of the other types. Experiments have steadily improved over the years. The evidence is strong that neutrino oscillations have not been observed from reactor sources. These negative findings do not contradict the positive indications; instead, they help pin down the magnitude of the masses and mixings that could explain all the experiments. Sensitive, long-baseline reactor experiments are under way in Europe, Japan, and the United States.

Neutrino oscillations are the most natural explanation for the solar neutrino puzzle described above. Several consistent experiments go into establishing the conclusion that too few electron neutrinos are coming from the Sun. The favored theoretical solution involves the exciting possibility that otherwise subtle effects of neutrino oscillations get amplified as neutrinos pass though the dense matter in the Sun. The evidence is strong, but the SNO experiments can confirm it conclusively by simultaneously measuring the oscillations of the electron neutrino and the total flux of all neutrino types. Using the Sun as a source for studying neutrino properties is a beautiful example of cross-disciplinary physics. Years of experimental and theoretical work to understand the nuclear-reaction power sources within the Sun and the thermodynamics and mechanical processes of the burning process provide the essential foundation for interpreting these results.

A recent experiment using the SuperKamiokande experiment in Japan is providing perhaps the strongest evidence for neutrino oscillations. For some

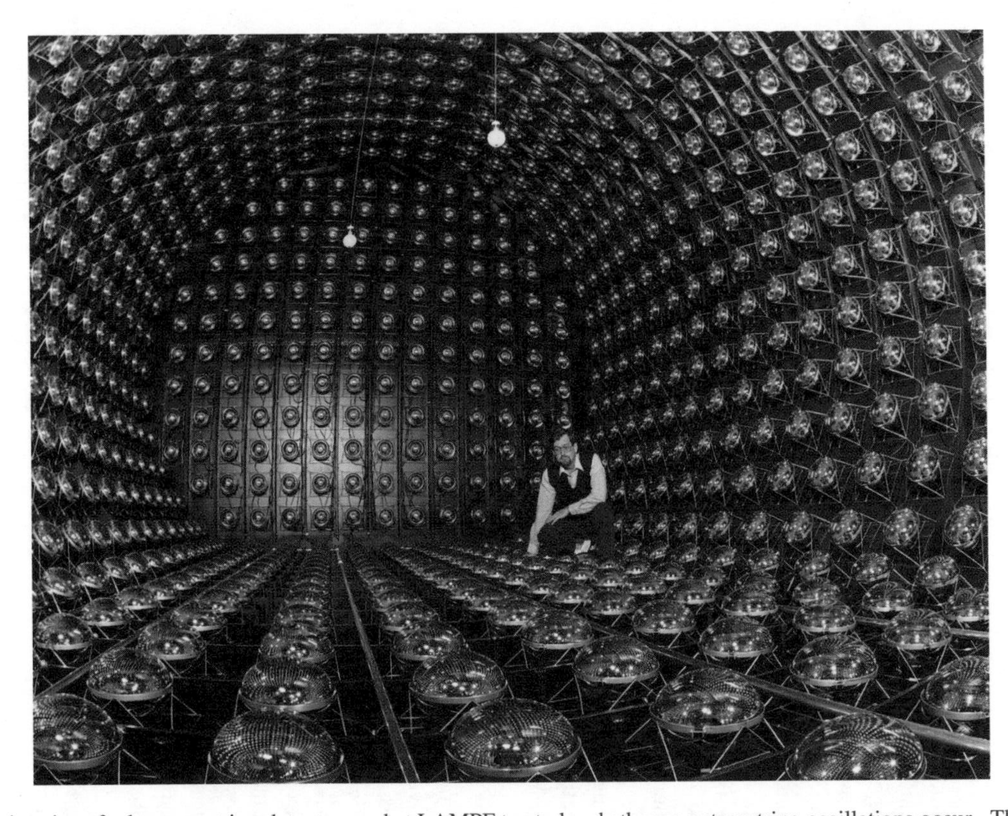

FIGURE 6.8 The interior of a large neutrino detector used at LAMPF to study whether or not neutrino oscillations occur. The active detector is a large tank of scintillating liquid viewed by about 1,000 large light-detecting photomultiplier tubes. This experiment provides evidence that could indicate that neutrinos do in fact have finite mass. (Courtesy of Los Alamos Meson Physics Facility.)

time, several large underground detectors indicated that the ratio of muon- and electron-type neutrinos detected underground was inconsistent with expectations based on a seemingly reliable estimate. The SuperKamiokande experiment has dramatically demonstrated that the anomaly can be traced to a difference in the number of upward going and downward going muon-type neutrinos that are detected. The conclusion is that muon-type neutrinos seem to be oscillating on their way to the detector, but that electron neutrinos do not oscillate by a detectable amount in that distance. That the electron-type neutrinos do not show evidence for oscillations here would be completely consistent with the conclusions from CHOOZ, the most recent and most sensitive long-baseline reactor experiment, and also with the solar results, indicating a rich structure of neutrino masses and mixing parameters in satisfying analogy to what is seen for the quarks. Many questions remain. Why are the neutrinos so light? Are neutrinos the same type of fermions as the charged leptons? Or, in the jargon of the field, Are they Majorana or Dirac particles?

Neutrinos with mass could have a dramatic effect on the extremely rare process of double beta decays. Double beta decay, with the emission of two electrons and two neutrinos, is simply the rare coincidence of two simultaneous beta decays. Two-neutrino double beta decay is expected to occur, even with massless neutrinos. Double beta decay was predicted early in the century, but it was directly observed only during the last decade. The first direct observation of this process was an experimental milestone made possible by some of the most advanced modern techniques of low background counting, invented specifically for this research. The observed double-beta-decay lifetimes are many trillions of times longer than the age of the universe. By now, two-neutrino double beta decay has been seen in about a dozen different isotopes. Double beta decays without neutrinos are also possible, if neutrinos have mass and are of the Majorana type, meaning that they are their own antiparticles. In this case, a neutrino emitted by one neutron can be absorbed by another, leaving two electrons that share all the decay energy. Double beta decay, with or without neutrinos, can be the dominant mode of decay for certain nuclear systems that are energetically forbidden to decay by ordinary beta decay. Neutrinoless double beta decay has not yet been observed, and whether or not neutrinos are of the Majorana type is still an open question, the answer to which will have profound implications. The current upper limit on the electron neutrino Majorana mass, derived from limits on neutrinoless double beta decay, is less than about 0.5 eV, and, thereby, experiments have already eliminated a number of previously proposed Standard Model extensions. The most sensitive experiment used several kilograms of isotopically enriched ^{76}Ge, which decays by the double-beta process. In this experiment, the ^{76}Ge itself was made into a detector capable of measuring the energy of decay electrons with excellent resolution. To avoid backgrounds, the experiment was done in an underground laboratory. Despite the great difficulties (the double-beta-decay process was observed only ten years ago for the first time), there has been tremendous progress in this field. There are now numerous observations in different systems. Experiments are being proposed that could improve existing limits on the neutrinoless process by orders of magnitude. Interpreting the limits in terms of fundamental neutrino parameters requires a close collaboration between the experimentalists and nuclear theorists.

THE WEAK INTERACTION WITHIN THE NUCLEAR ENVIRONMENT

The nuclear interior provides a unique environment for studying the weak interaction between hadrons, with no leptons involved. The weak force is exceedingly feeble compared to the strong nuclear force, but parity violation provides us with the necessary clue for recognizing its effects. While the weak force can mostly be ignored in describing the structure of nuclear systems, it still introduces a small violation of mirror-reflection symmetry. The weak interaction in its purest form has too short a range to act directly; instead, it organizes itself as a perturbation to the various meson exchange forces between the nucleons. Much has been learned about the weak hadronic interaction, but significant puzzles remain to be solved. Sensitive experiments combined with sophisticated theory are still inadequate to provide a completely consistent picture of the structure of the weak force within the nucleus. Investigations of the weak-interaction-induced parity mixing of nuclear levels in well-chosen systems can discriminate between the charged current (arising from *W* exchange) and neutral current (arising from *Z* exchange) components of the nucleon-nucleon weak force. The effect of the charged currents is consistent with expectations, but there is no clear evidence of the expected neutral current effect in nuclei. However, recent evidence of a nuclear anapole moment (basically a doughnut-shaped or toroidal nuclear magnetic field configuration within a nucleus) from studies of atomic parity mixing is consistent with the expected size of the nucleon-nucleon neutral current. New experiments are clearly called for. Although the effects are extremely small and hard to observe in the nucleon-nucleon system, they have been seen directly in proton-proton scattering. A new experiment is planned to measure the effect of parity violation on the gamma rays following the capture of polarized neutrons on protons. This experiment could solve the puzzle about the missing effect of the *Z* in nuclei, because it is free from the theoretical uncertainties that are required to interpret finite nuclei. However, the experiment is exceedingly difficult, because the size of the expected effect is only at the part in 100 million level.

The weak force should also affect the outcome of neutron-nucleus elastic-scattering experiments. These experiments are sensitive to the interference of low-energy neutron capture resonances with angular momenta of zero and one. The method is exquisitely sensitive to a small parity-breaking force, because the nucleus acts as an amplifier. In experiments with polarized neutron beams from spallation sources, parity-mixing effects as large as several percent have been observed. Analyzing the observed effects in nuclei is an example of the interplay between nuclear structure and the fundamental interactions. Statistical models of the nuclear state distributions are used to extract the mean values of the parity-mixing matrix elements. Figure 6.9 shows some of the data from a measurement of parity-violating neutron scattering on thorium. The size of the parity-violating effects indicates dramatically how the nucleus can be used to amplify small features of the fundamental forces.

EXOTIC PARTICLE SEARCHES, RARE DECAYS, AND NUCLEAR PHYSICS

Nuclear systems are often exploited in attempts to verify speculations about hypothetical new particles. For example, nuclear experiments were instrumental in studying the possible existence of a particle called the axion, proposed several

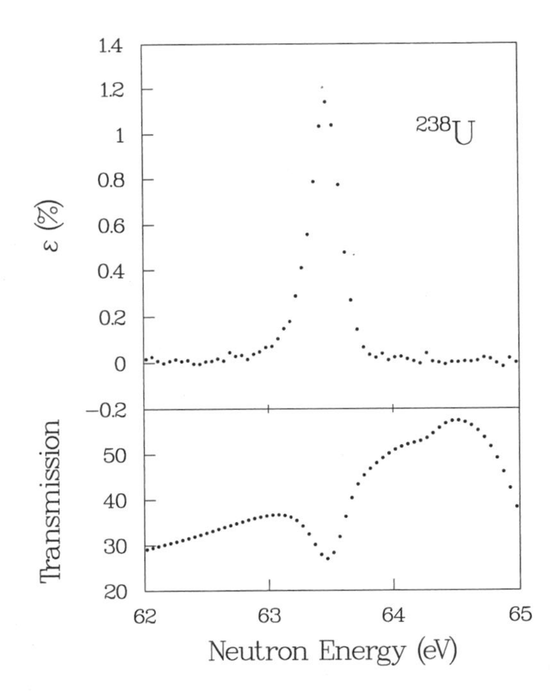

FIGURE 6.9 Data from an experiment measuring the transmission of polarized neutrons through a target of ^{238}U. The difference in transmission for the two neutron polarization states demonstrates parity violation in the nuclear force. The enormous effect, larger than a percent, is a consequence of the nucleus's ability to amplify an exceedingly small intrinsic parity violation from the weak interaction.

years ago as a consequence of a new symmetry introduced to explain the observed CP symmetry of the strong interaction. As first proposed, the axion would decay to two photons or positron-electron pairs. Over the course of several years, this possibility was ruled out by a series of searches, primarily using nuclei. Unfortunately, the mass of the axion and its interaction strengths are not completely specified by theory, so experiments have not yet excluded them completely. An ongoing experiment searches for very light axions. This experiment exploits the predictions that axions couple to photons and that very light ones should have an effect on a finely tuned electromagnetic resonance cavity.

Nuclear techniques have been applied to searches for free quarks, very light scalar particles, and several other possible particles suggested by various theories. Nuclear physicists have been quick to apply their sizable arsenal of experimental techniques to discovering whether or not particular proposals concerning new particles correspond to what actually exists in nature.

Decay mechanisms that are forbidden within the context of the Standard Model might otherwise be expected. For example, the proton should be stable in the Standard Model, but it is expected to have a decay mode in various grand unified models, which extend the Standard Model by including quarks and leptons in more democratic multiplets. Proton-decay experiments have mostly been undertaken by high-energy physicists, with some participation from the nuclear physics community. These experiments are being steadily improved, and the upper limit on the proton lifetime is now longer than about 10^{32} years. A related process of neutron-antineutron oscillations was also an active area of investigation during the last decade. Again, no effects were seen, but useful limits were obtained from experiment. Observing forbidden decays of mesons and leptons might also be signals of physics beyond the Standard Model. One possibility that has received much attention is the search for the decay of the muon into an electron and a gamma ray. Several attempts to find this process have been made at medium-energy accelerators around the world. These limits, like the limits on the neutron and electron electric-dipole moment, provide some of the best constraints on supersymmetric extensions to the Standard Model.

OUTLOOK

This area of research is motivated by a desire to gain new insights into the most fundamental aspects of the structure that underlies nuclear physics. It has remained a vital subfield of nuclear physics because the issues are fundamental and because a constant flow of new experimental opportunities has been made possible by capitalizing on advances in experimental physics. Advancements in the fundamental experimental tools of nuclear physics—the newest accelerators and the most advanced detector systems—are essential for the future health of this research. Ironically this research area also relies on some of the oldest accelerator facilities of nuclear physics. Much of the work is carried out at small accelerators at national laboratories and universities, small tandem accelerators, and cyclotrons. Access to small facilities with specific capabilities for particular experiments is also an important priority for this subfield. This area enriches the field of nuclear physics by sharing frontiers with other fields of physics, overlapping and enhancing work in atomic physics, high-energy particle physics, and astrophysics.

Although nearly all new facilities for nuclear physics will be exploited for testing the Standard Model, some future facilities have obvious utility that make them a high priority for the field. The new generation of intense sources of cold and ultracold neutrons hold the potential of allowing the Standard Model's version of the weak interaction to be better tested in a simple hadronic system. Precision studies of neutron beta decay and better searches for a neutron electric-dipole moment require access to the most intense research reactors and spallation sources available. Access to more intense neutron sources that become available in the next decade has a high probability of leading to a major discovery in this area of research.

The venerable area of nuclear beta decay has a new life because of the advent of new techniques of atom manipulation, new methods of polarizing nuclei, sophisticated online isotope separators, and new radiation detection methods. Ion or atom traps offer new opportunities for reducing the systematic uncertainties that now limit this research, but these techniques make more stringent demands on the production of particular radioactive species. A high-intensity exotic-beam accelerator would also be the source of radioactive atoms required for this research, and an exotic-beam facility is a high priority for this area of research.

In the next five years the newest solar neutrino experiments should finally resolve the question of whether or not the neutrino and its interactions are the cause of the solar neutrino problem. The discovery of neutrino masses and mixing could lead to the next scientific revolution, leading to the first significant change in the Standard Model two-and-a-half decades after it was established. Currently there are two other experimental indications that neutrinos may have mass. The signal from atmospheric neutrinos should be elucidated in a series of planned, long-baseline experiments at accelerators. The second indication appeared in a medium-energy nuclear physics accelerator experiment. Nuclear physicists will continue to participate in the follow-up experiments that should resolve these issues in the coming years.

7

The Tools of Nuclear Physics

INTRODUCTION

Nuclear physics is a science driven by experiment, so its progress depends critically on advances in instrumentation. New developments in accelerators, detectors and their associated electronics, data acquisition systems, and computers for data analysis have been the bases for rapid developments in the field and provide the technical underpinnings for today's thrusts in nuclear physics.

Given the range of nuclear phenomena, and the corresponding length and energy scales at which we observe them, a large variety of experimental instrumentation is needed and is continuously being developed. This spectrum includes various types of accelerators that provide a broad range of beam species and energies, as well as the particle detectors to register and identify a diverse range of reaction products.

Beams of nucleons and nuclei from low-energy accelerators are used to excite the nucleus as a whole. At the other end of the energy scale, high-energy lepton and hadron beams probe the distance scales relevant for the subnucleonic structures of nucleons and nuclei. Colliding beams of massive, highly relativistic nuclei transiently generate nuclear matter at the highest energy densities and hadron densities, offering the possibility of observing the quark-gluon plasma. In addition, exploring spin degrees of freedom requires polarized beams (and targets), while beams of mesons and strange baryons uniquely probe specific aspects of the strong interaction in nuclear many-body systems.

To the same extent that a diverse assembly of accelerators with widely varying characteristics is needed to cover the range of beam species and energies needed for nuclear physics, detectors of greatly varying characteristics have been and continue to be developed for a spectrum of experimental observables. These range from the high-resolution detection of optical radiation at eV energies in studies of nuclear hyperfine structures, to sophisticated multidetector arrays needed to disentangle thousands of reaction products in high-energy nuclear collisions, and to scintillation detectors of thousands of tons buried deep underground to register the most elusive particles in nature, the neutrinos coming from the cosmos, the Sun, or from accelerators. Detectors are used in stand-alone mode, as in the underground neutrino experiments; as single detectors and small arrays at low-energy facilities; and in vast assemblies of complex particle detection systems at the facilities with the highest-energy beams.

The continuous advancement of accelerators, detectors, and data acquisition techniques provides a rich milieu for training and innovation over a wide spectrum of technical areas, including electronics, vacuum technology, large-scale data acquisition and computer systems with corresponding software development, novel detector materials and sensors, automated high-level control systems, ion-beam and accelerator technology, and superconductivity.

The experimental work in nuclear physics goes hand in hand with the development of theoretical understanding. The theoretical effort is undertaken by a number of researchers at universities and laboratories. An important part of the infrastructure of the field is the Institute of Nuclear Theory, described in Box 7.1.

ACCELERATORS

Nuclear physics needs primary beams of electrons, protons, and heavy ions over a wide energy range. Each serves a complementary class of experiments. Secondary beams of other particles, such as neutrons, pions, muons, neutrinos, and radioactive ions, can be derived principally from intense proton and heavy-ion beams. Accelerator facilities for nuclear science fall into two major categories: larger facilities that operate for substantial outside-user communities, and smaller facilities that mainly serve local groups of scientists.

Historical Perspective

A variety of accelerators was invented, built, and used in the 1930s to begin the exploration of nuclei. During the 1950s and 1960s, needs from nuclear physics experiments led to major improvements in these technologies and a number of accelerators, cyclotrons, and Van de Graaffs (now considered small) were constructed and used for research at university laboratories. As the requirements for beam energies, intensities, and especially beam species grew, larger dedicated facilities were built at universities and at the national laboratories. It was also during this period that high-energy physics started, and the accelerator developments of nuclear physics formed the basis of the first high-energy facilities—

BOX 7.1 The Institute for Nuclear Theory

The formulation of our theoretical understanding of nature and frequent, detailed discussions between theorists and experimentalists, are key elements to progress in physics. About a decade ago, with CEBAF and RHIC on the horizon, the Nuclear Science Advisory Committee recommended establishing a center for nuclear theory. As a result, the national Institute for Nuclear Theory (INT) opened its doors in the spring of 1990 at the University of Washington in Seattle.

Some of the goals of the INT are as follows:

- Create a research environment where visiting scientists can focus their energies on key frontier areas;
- Encourage interdisciplinary research at the intersections of nuclear physics with related disciplines, such as particle physics, astrophysics, atomic physics, and condensed matter physics;
- Recruit and nurture the best young researchers;
- Contribute to scientific education through INT schools;
- Strengthen international cooperation in nuclear physics and physics generally, through cooperative programs and exchanges.

The INT contributes in important ways to the vitality of nuclear physics. It sponsors three programs each year focused on specific physics questions and draws visiting physicists from the United States and abroad who are expert in the subject area. The INT also sponsors an annual series of schools and smaller workshops (e.g., see Figure 7.1.1), many of the latter organized quickly in response to urgent developments in the field.

The INT has enjoyed remarkable success from its inception. The three main programs attract approximately 200 visitors each year for an average stay of about one month. Approximately 60 percent of the visitors are U.S. physicists; the rest come from around the world. The concept behind the programs is that close discussions among groups of experts will both sharpen the physics questions and speed progress on their resolution. This includes issues specifically related to experiment: most programs include significant participation by experimentalists, who describe what can be measured and how theory might be of help in their analyses. Many of the more interdisciplinary programs have drawn equal numbers of participants from several physics subdisciplines. Recent examples of such programs include nucleosynthesis, atomic clusters, Monte Carlo techniques in strong-

greatly aided by the discovery of strong focusing. In nuclear physics the Los Alamos Meson Physics Facility (LAMPF) was the premier large nuclear physics facility in the United States for many years. At the same time, innovative technological developments and extensions of existing smaller accelerators—superconducting magnet developments, post accelerators based on superconducting radiofrequency technology, ion sources for highly charged ions, intense polarized

FIGURE 7.1.1 A group of students in front of the Institute for Nuclear Theory, participating in INT's Research Experiences for Undergraduates program.

ly interacting many-body systems, and connections between high-energy and low-energy precision measurements of the weak interaction.

The INT workshops, which typically run from 2 to 5 days, attract another 200 visitors each year. Two of the workshops are an ongoing annual series cosponsored by CEBAF and RHIC. They have as their focus the theory needs associated with specific experiments. Other occasional workshops have involved collaborative efforts with INT's counterpart in atomic physics, the Institute for Theoretical Atomic and Molecular Physics (ITAMP) at Harvard, and with overseas institutes. Still others are sponsored entirely by INT in response to urgent physics questions. A recent example was a workshop on the nuclear physics of the pp chain, crucial to the interpretation of solar neutrino experiments. The INT helps administer the National Summer School in Nuclear Physics, an annual school for advanced graduate students.

electron and proton beams, storage rings for leptons and hadrons—have provided new and unparalleled capabilities for nuclear physics research. In addition, high-energy physics facilities were used in certain areas of nuclear physics. For example, the Lawrence Berkeley National Laboratory (LBNL) Bevatron was modified to become the Bevalac, the first high-energy heavy-ion accelerator, and nuclear physicists carried out a number of experiments at the Stanford Linear

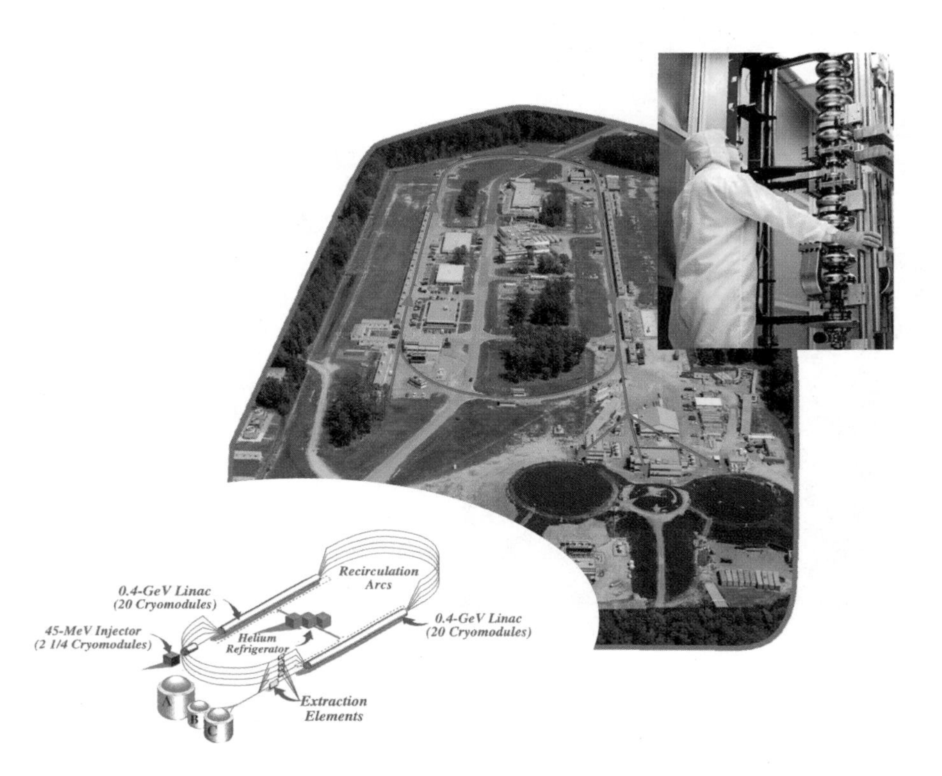

FIGURE 7.1 Schematic and aerial view of CEBAF at the Thomas Jefferson National Accelerator Facility (TJNAF) in Newport News, Virginia. The accelerator is based on superconducting cavities for acceleration. Beams are recirculated and accelerated up to 5 times to provide intense continuous electron beams of 1-5 GeV, polarized and unpolarized, to three target areas simultaneously. The photo insert shows assembly of a superconducting cavity section in the clean room. (Courtesy TJNAF.)

Accelerator Center (SLAC), and continue to be involved in measurements at Fermilab and other high-energy facilities around the world.

Two current key objectives of nuclear physics, understanding the quark structure of nucleons and nuclei and the search for the quark-gluon plasma, have demanded higher energies, beam currents, and duty factors for electrons, and much higher energies for heavy ions. This motivated the construction of two large facilities designed specifically for nuclear physics at these research frontiers: CEBAF at the Thomas Jefferson National Accelerator Facility (TJNAF) in Virginia (shown in Figure 7.1), a high-current, continuous-beam electron accelerator of moderately high energy that has recently become operational; and the

Relativistic Heavy Ion Collider (RHIC), the first high-energy heavy-ion collider, now nearing completion at Brookhaven National Laboratory (BNL).

Accelerator Research and Development

Most of the accelerator facilities in nuclear physics have developed in close association with the science on the one hand and with the more general research and development in accelerator science and technology on the other. Long-term accelerator R&D has led to new technologies and new science in nuclear physics, with significant spin-off into other areas of science.

Superconducting magnet development at Michigan State University has led to compact, powerful cyclotrons for heavy-ion beams; superconducting beam bending and focusing magnets for the highest-energy beams, originally developed for proton accelerators in particle physics, have been extended for use with heavy-ion beams at RHIC.

Superconducting radio-frequency (rf) cavities have been successfully developed over the past two decades for acceleration of low-velocity heavy ions and of high-velocity electrons. Cavities for heavy ions, first developed and implemented at Argonne National Laboratory in the early 1970s and soon after at the State University of New York at Stony Brook and other universities, provide cost-effective and versatile postaccelerators for high-quality ion beams that cover the full mass range from protons to uranium. The successful operating experience with this new technology bolstered its development and application to the higher velocities required in the accelerators for particle physics. In nuclear physics, recent R&D at TJNAF led to the development of reliable, high-gradient, superconducting acceleration cavities with higher-order mode suppression for electrons at the velocity of light. R&D in this area could lead to a further factor of two or so in gradient and will likely open this technology to other applications, such as the use of superconducting rf cavities as accelerator drivers for high-power interaction regions and ultraviolet free-electron lasers and, possibly, in high-energy linear colliders.

The development of high-charge-state electron cyclotron resonance (ECR) ion sources is revolutionizing the acceleration of heavy-ion beams both at cyclotrons and at heavy-ion linacs. Since energy gains increase with ionic charge state, acceleration of highly charged ions makes the most efficient use of accelerator voltages.

The development of polarized beams and targets has made major strides. Developments in atomic-beam techniques at the University of Wisconsin are providing polarized gas targets and polarized ion sources, and long-lived frozen polarized targets have been developed at various facilities. Laser polarized targets have been developed for hydrogen, helium, and lithium ions at various universities and national laboratories. So-called Siberian snakes and spin rotators

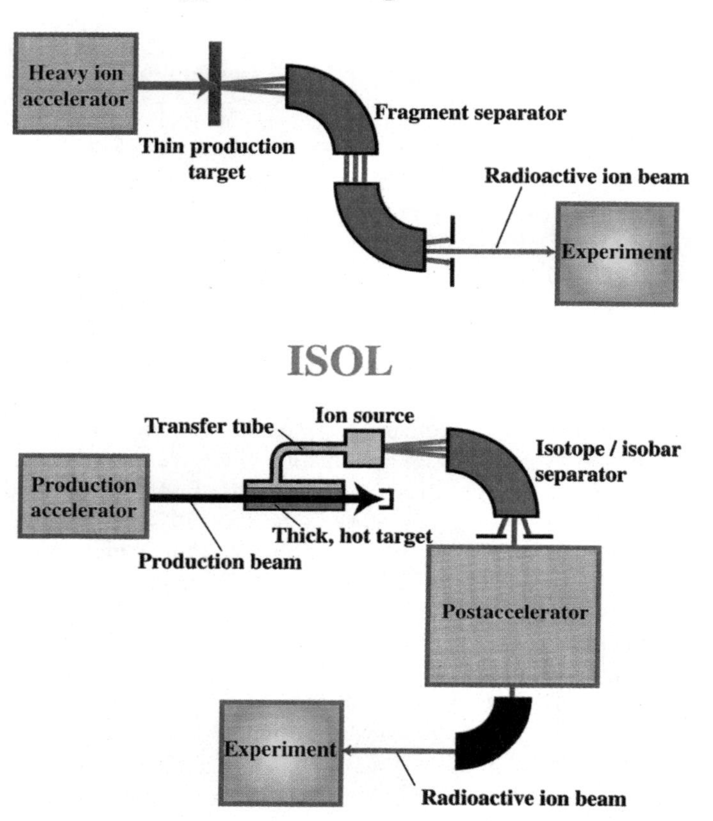

FIGURE 7.2 Schematics of the two principal concepts for the generation of energetic beams of shortlived nuclei (radioactive beams). The top figure illustrates the in-flight projectile fragmentation approach in which an energetic particle (typically several tens of MeV/u to GeV/u) is fragmented in a nuclear reaction in a thin target, and radioactive reaction products are separated in flight and transported as a secondary beam to the experiment. The ISOL (isotope separator online) approach illustrated at the bottom employs two independent accelerators: a high-power driver accelerator for production of the short-lived nuclei in a thick target that is directly connected to an ion-source and a postaccelerator. Radioactive atoms diffuse out of the hot target into the ion source where they are ionized for acceleration in the postaccelerator. This scheme provides for excellent beam quality since it allows full control over beam energy and time structure.

based on a superconducting helical dipole magnet design are allowing acceleration and complex manipulations of polarized beams.

Electron cooling of ion beams in storage rings has greatly advanced, allowing precision in-beam nuclear physics measurements in storage rings with internal targets. The low emittance of the cooled, stored proton beam at the Indiana University Cyclotron Facility (IUCF) has proved to be an excellent tool to study threshold reactions for nuclear physics on the one hand and nonlinear, single-particle beam dynamics effects of interest to accelerator physicists on the other.

Some of the developments mentioned have occurred at the larger user facilities. But often smaller facilities, in particular those at universities, have provided the environment for generic developments that have important applications to the research programs at large accelerators. This exemplifies the symbiotic relationship between small university accelerators and large national and international facilities in nuclear physics.

Several new projects are in progress or under consideration. First, there is the completion of RHIC, expected for 1999, with ongoing R&D efforts to extend its performance beyond the design parameters: superconducting magnet control, bunched-beam stochastic cooling, and the implementation of polarized proton-beam operation. A viable, bunched-beam, stochastic cooling system could potentially ameliorate the effects of intrabeam scattering in heavy-ion beams and thus significantly improve performance. A recent decision to implement polarized proton operation at RHIC has led to an R&D program to develop the previously mentioned Siberian snakes and spin rotators.

The 1996 NSAC *Long Range Plan* recommended an upgrade, now under way, of the radioactive-beam capabilities at Michigan State University. Two superconducting cyclotrons are being upgraded to boost the energy and intensity of the primary production beam at this radioactive-beam facility of the in-flight fragmentation type. Radioactive-beam intensities will be increased by factors ranging from 100 to 1,000. Figure 7.2 shows the schematics of the two principal approaches to radioactive-beam facilities—in-flight fragmentation and the two-accelerator ISOL (isotope separator online) concept.

The NSAC *Long-Range Plan* also recommends development of a next-generation, ISOL-type, radioactive-beam facility based on the two-accelerator (driver/post-accelerator) concept, with its construction to begin once construction of RHIC is substantially completed. The advanced ISOL facility will bring together the fruits of a number of current accelerator developments—suitable radiofrequency quadrupole accelerators, high-power conventional and superconducting linacs or cyclotrons, and the developments of target/ion source systems—that have been carried out in recent years at a variety of nuclear physics facilities.

A list of present accelerators for nuclear physics research in the United States is given in Appendix A with a short summary of their respective performance characteristics.

INSTRUMENTATION

Major progress made in target and detector systems in recent years underlies many of the current advances in nuclear science. Polarized targets of light nuclei, produced either by atomic-beam techniques or by novel approaches using laser-induced polarization, have played an important role in fixed-target experiments, as well as in experiments using internal targets in storage rings. Advances in particle detectors range from single detectors at low energies to complex detector arrays for the highest energies and the largest particle multiplicities. Most detectors are used in experiments with beams from accelerators, but stand-alone systems also play an important role and have undergone major developments. It is not possible to give a detailed discussion of all the recent developments within the limited space available in this report. Rather, we will give selected examples from the various areas of nuclear physics research to provide a taste of the novel developments and genuine advances that these targets and detectors represent for the science.

Examples of New Detector Systems

Ion and Atom Traps

At the lowest energies, novel structure information on nuclei has been obtained by trapping nuclei, in particular short-lived, radioactive ones far from stability, in ion or atom traps and by measuring their masses and the properties of their radioactive decays.

Trapping atoms in magneto-optical traps with the help of lasers has made major advances over the last decade (as shown in Figure 7.3). This technique promises unique advances in studying nuclear decays, allowing precision studies of fundamental symmetries under controlled conditions. For example, atomic experiments can measure parity-violating matrix elements with significant precision. The heaviest alkali, francium, promises particularly large sensitivity to parity violations because of its simple structure, with only one electron outside of a spherical core. Francium has no stable isotopes, making it difficult to carry out this important class of measurements. However, several laboratories have recently taken an important step in starting these measurements, by efficiently collecting the small number of francium atoms created in a nuclear reaction and cooling them sufficiently to confine them in a laser trap. To reach the intensities of trapped ions needed for the symmetry studies, major advances are needed in production yield and trapping. The next-generation ISOL facility will provide the required intensities.

The use of ion traps in nuclear physics has proven equally exciting. Penning traps have been effective instruments for high-accuracy mass determinations. With these devices, the mass ratios of a number of stable or long-lived light

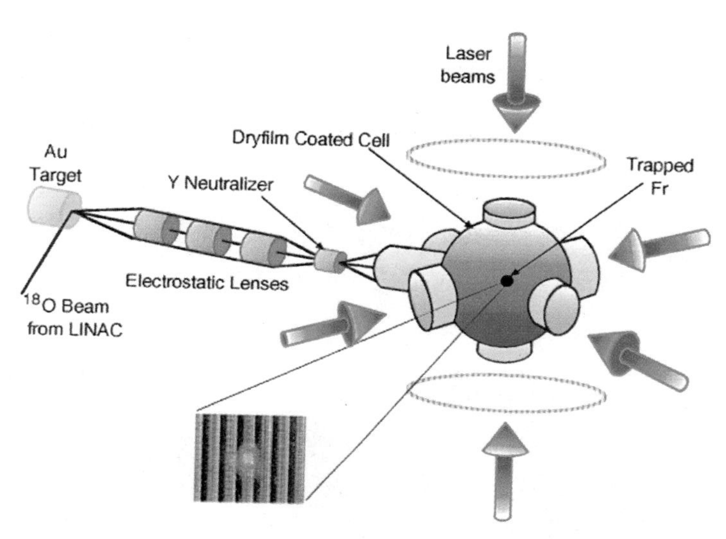

FIGURE 7.3 Schematic layout of a magneto-optical trap, which uses a combination of a weak magnetic field and six intersecting polarized laser beams to capture radioactive atoms generated in a nuclear reaction. These atoms are captured by neutralization and thermalization of a low-energy radioactive ion beam. Once captured in the trap, these atoms can become an ideal source for precision studies of nuclear decays that address aspects of fundamental symmetries and the Standard Model.

particles have been determined to high accuracy, on the order of a part in a hundred million. High-accuracy measurements with radionuclides have been performed for an accurate determination of the binding energies of nuclei.

Exploring the Structure of Exotic Nuclei

Novel high-precision mass measurements have also been made in a storage ring. Signals have been detected from as little as a single circulating ion of a specific isotope.

Advanced detectors for nuclear reactions near the Coulomb barrier allow one to explore the behavior of the nucleus under the extremes of excitation, spin, and neutron-to-proton ratio. High-sensitivity recoil systems allow identification of the rarest reaction products in heavy-ion fusion reactions at zero degrees, which are usually hidden in an overwhelming background from the incident particle beam. This has allowed nuclear physics, for example, to identify the new heavy elements up to 112, and a number of proton emitters, literally dripping protons, at the proton drip line.

New information about high-spin states in nuclei has been obtained with large, highly efficient germanium detector arrays. The Gammasphere array consists of 100 large Compton-suppressed germanium crystals. Using its high efficiency for high-energy gamma rays, it has been possible to identify, unambiguously, the transitions linking states in the superdeformed and normally deformed potential minima. The combination of such powerful high-resolution detectors with the highly selective recoil systems shown in Figure 7.4, supports studies of nuclear structure out to the limits of nuclear existence. Similarly, new high-resolution magnetic spectrometers allow precision nuclear structure studies in reactions with radioactive beams or polarized light-ion beams. Figure 7.5 shows the recently completed superconducting magnetic spectrograph at Michigan State University for a broad range of studies with radioactive beams.

FIGURE 7.4 Gammasphere, the national gamma-ray facility for studies in nuclear structure research, mounted in front of the fragment mass analyzer (FMA) at the ATLAS accelerator at Argonne National Laboratory (ANL). Gammasphere was built at the Lawrence Berkeley National Laboratory in a collaboration of national laboratories and universities and is the premier detector for nuclear structure research with gamma rays. The FMA is an electromagnetic ion-beam filter that allows researchers to select rare nuclear reaction products out of an overwhelming background of beam particles at zero degrees, and then study their decay through coincidence with the cascade of decay gamma rays detected by Gammasphere. (Courtesy ANL.)

FIGURE 7.5 The superconducting magnetic spectrograph S800 at Michigan State University. The target for the nuclear reactions is at the bottom left of the spectrometer structure. Reaction products from nuclear collisions, induced by beams of shortlived nuclei (radioactive beams) are selected and bent upwards by powerful superconducting magnets onto the focal-plane detector system mounted on the top of the spectrograph. (Courtesy MSU.)

Hall A

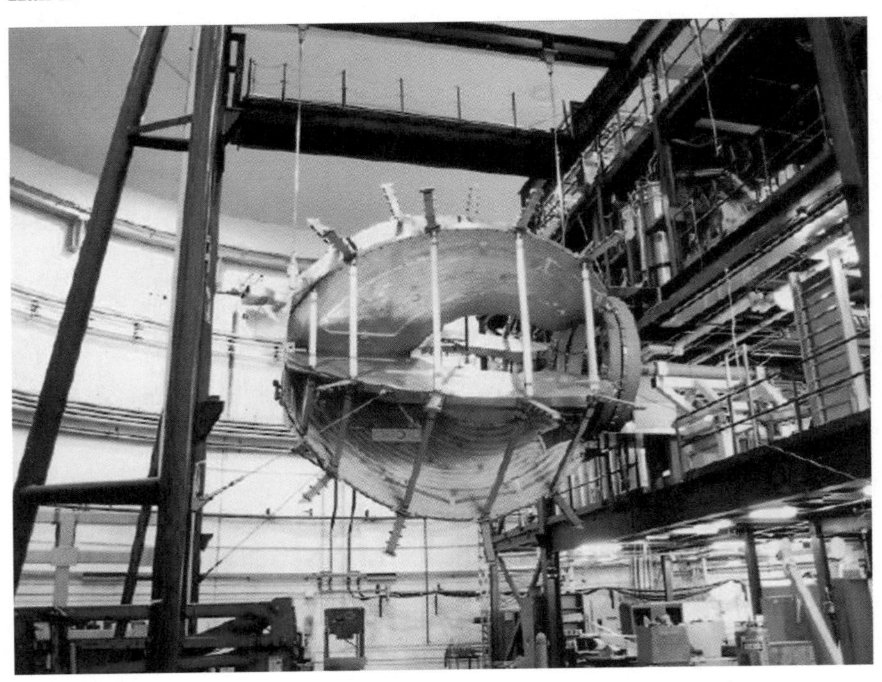

Hall B

Hall C

FIGURE 7.6 The three experimental halls at CEBAF at the Thomas Jefferson National Accelerator Facility in Newport News, Virginia. Hall A houses two high-resolution magnetic spectrometers based on superconducting magnetic dipoles to bend the reaction products into their focal-plane detector systems. Hall B houses a clam-shell-type, large-acceptance, magnetic spectrometer, again based on superconducting magnets, but with a novel technology that uses iron-free magnet coils. Hall C houses a combination of a high-resolution magnetic spectrometer for electrons and a short-orbit spectrometer to allow detection of short-lived particles before they decay. (Courtesy TJNAF.)

Detectors for the Quark Structure of Matter

The program to elucidate the quark and gluon substructure of nucleons and nuclei requires large, high-quality spectrometers at high energies and momenta. At TJNAF, two of the three experimental areas house pairs of large magnetic spectrometers (shown in Figure 7.6). The spectrometers are equipped with sophisticated focal-plane detection systems with high rate capabilities, to detect, identify, and analyze with excellent resolution electrons as well as hadronic reaction products, i.e., mesons, nucleons, and hyperons. A third hall houses a novel large-acceptance spectrometer, with its magnetic field generated by a six-coil, iron-free, superconducting toroidal magnet and a complex particle detector array. At MIT's Bates accelerator, a novel detector, the Bates Large Acceptance

Spectrometer Toroid (BLAST), is under construction for a full exploitation of spin observables in intermediate-energy nuclear physics. It will be used in experiments in which polarized circulating electrons scatter from polarized internal targets in the Bates storage ring. Out-of-plane measurements, for added understanding of electron scattering from nuclei, are being performed with the new Out-of-Plane Spectrometer (OOPS), shown in Figure 7.7.

The detection of neutral mesons with high resolution and efficiency has been a long-standing technological challenge. Arrays of barium fluoride detectors have been developed for use in heavy-ion-induced reactions, where excellent timing characteristics are necessary to discriminate against the overwhelming neutron background. A detector for high-resolution measurements of pions and eta mesons that detects high-energy decay photons with bismuth germanate (BGO) converters and pure cesium iodide (CsI) calorimeters was developed at Los Alamos and recently moved to the Brookhaven accelerators.

Detectors for the Frontier of High-Energy Density

The attempt to observe phase transitions in bulk nuclear matter by colliding two heavy nuclei at RHIC requires detectors covering the full solid angle, having large dynamic range, and, at the highest energies, able to handle the highest particle densities emerging from any reaction studied in the laboratory. At medium energies, the clean detection and identification of complex nuclear fragments pose severe challenges. The RHIC detectors collectively cover all of the predicted signatures for the quark-gluon plasma (QGP). Two complementary major detector systems, PHENIX and STAR, and two smaller-scale detector systems, PHOBOS and BRAHMS, are under construction.

The STAR detector, shown in Figure 7.8, uses the large, solid-angle tracking and particle identification capability of a cylindrical time-projection chamber, placed in a large solenoidal magnet. The design emphasizes detection of the global features of the hadrons and jets as the signatures for QGP formation. Proposed additions to this detector include a silicon vertex tracker, an electromagnetic calorimeter, a time-of-flight array, and a pair of external time-projection chambers.

The PHENIX detector focuses on the detection of leptons, photons, and hadrons in selected solid angles, with a high-rate capability, to emphasize the electromagnetic signatures of QGP formation. The central part of PHENIX consists of an axial field magnet and two detector arms, each covering one-fourth of the full azimuth. Each arm will be equipped with a drift chamber, pad chambers, ring imaging Cerenkov counter, time-expansion chamber, time-of-flight, and electromagnetic calorimeters. Silicon detectors close to the beam pipe provide nearly full solid-angle coverage for particle detection. Upgrade items proposed for PHENIX are instrumentation for the forward muon arm, a high-resolution photon detector, and improved tracking and trigger electronics.

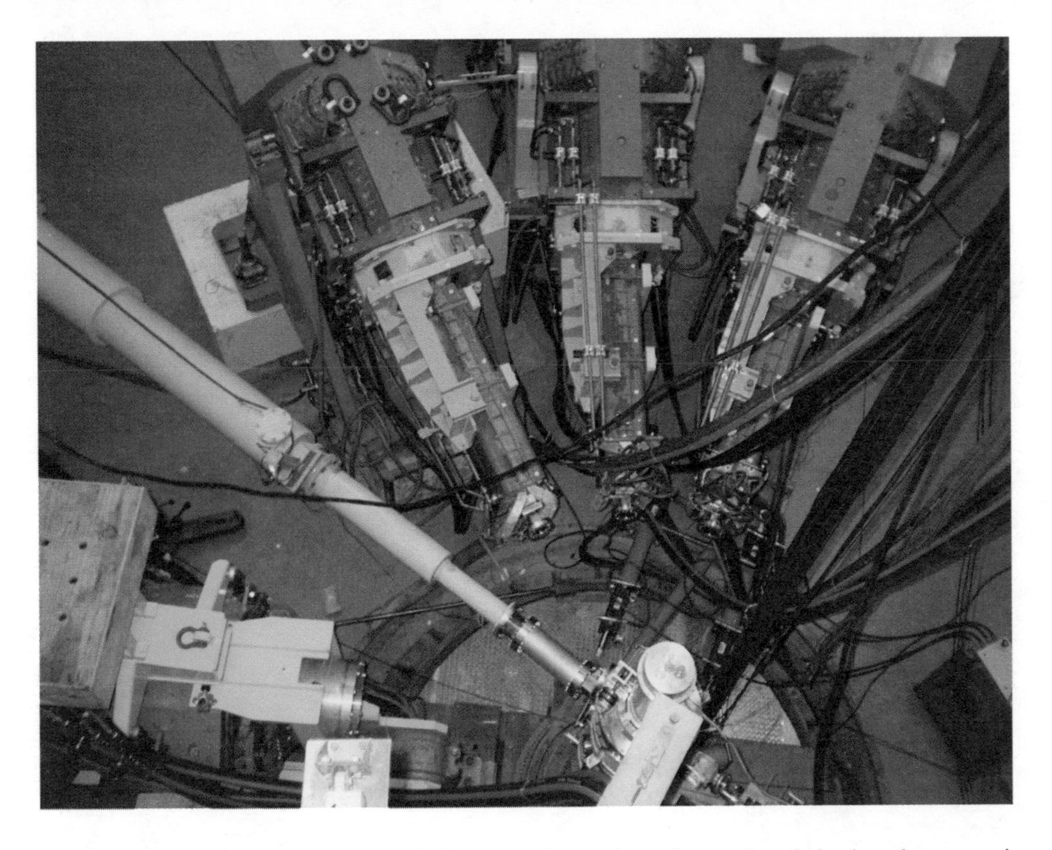

FIGURE 7.7 The Out-of-Plane-Spectrometer (OOPS) at the Bates accelerator is used to study polarization phenomena in nuclei. The four spectrometers allow positioning out of the reaction plane, which is important for the analysis of polarization variables. (Courtesy MIT Bates Linear Accelerator Center.)

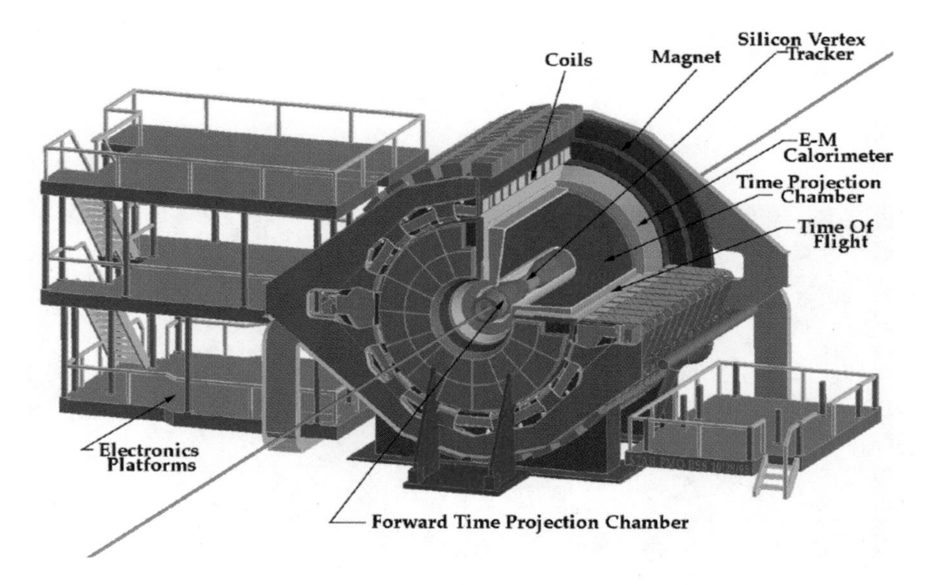

FIGURE 7.8 The STAR detector at the Relativistic Heavy Ion Collider (RHIC) at Brookhaven National Laboratory. This detector comprises an assembly of many different detection systems, including a large central time-projection chamber that allows precise three-dimensional reconstruction of particle trajectories. Appendix Figure A.2 shows an aerial view of the RHIC complex. (Courtesy RHIC.)

PHOBOS is a smaller detector; its double arm spectrometers use small magnets with strong magnetic fields and high-spatial-resolution silicon detector planes. The detector focuses on hadronic signatures for the QGP at low transverse momentum. BRAHMS uses two independent magnetic spectrometer arms for inclusive measurement, with particle identification and the ability to reach very forward angles.

Detection Schemes for Fundamental Symmetries and Underground Laboratories

High-precision spectrometers provide measurements of endpoints for beta-decay spectra that limit the electron-neutrino mass to a few eV. A neutrino-oscillation experiment at Los Alamos is providing tantalizing glimpses of a possible oscillation between different neutrino flavors, using a 200-ton liquid scintillator to detect both scintillation and Cerenkov light and identify the different neutrino-interaction processes. Two smaller scintillation detectors to search for oscillations using reactor antineutrinos have been constructed in California and in France.

Underground laboratories are utilized increasingly to solve specific problems in nuclear physics; they have been essential in some important major advances of the field, particularly for measurements requiring very low backgrounds. They are cleaned of contaminants and are well shielded from cosmic rays, thus providing the environment in which searches for incredibly rare events become possible. For example, improved studies of the elusive double-beta decays established important upper limits on the probability for neutrinoless decay. Three new, underground, solar-neutrino detectors (and one extension of a previous detector) are under construction or recently completed; these will generate large amounts of precise data over the next few years that will have a major impact on understanding how the Sun shines and on the nature of neutrinos. Following up on the tradition of having pioneered this field (with the chlorine experiment in the Homestake mine), U.S. nuclear physicists are participating in all these experiments.

The Kamiokande detector, a large tank of water in which the Cerenkov light induced by charged particles is detected with great sensitivity, was built under a mountain west of Tokyo to study proton decay. Proton decay was not observed, but this detector became the first to be able to count individual neutrino events from the Sun. Together with the IMB (Irvine, Michigan, and Brookhaven) detector, Kamiokande also recorded the burst of neutrinos from a supernova explosion in 1987. Recently the international team of nuclear and particle physicists working at the new SuperKamiokande detector announced the discovery of oscillations in neutrinos produced at the edge of outer space from the interaction of cosmic-ray particles with air molecules. The consequences for physics are likely to be revolutionary.

The Sudbury Neutrino Observatory (SNO) is a water-Cerenkov detector like SuperKamiokande; it is sited 2 km underground near Sudbury, Ontario, in an active nickel mine. Unlike SuperKamiokande, SNO contains heavy water. Although SNO will have a broad range of physics and astrophysics capabilities, its primary activity will be the detection of neutrinos from ^{8}B decays in the Sun. With its deuterium nuclei, SNO will be able to distinguish electron neutrinos from the other neutrino varieties.

The Gran Sasso Laboratory, adjacent to the highway tunnel under the Gran Sasso d'Italia, in the Apennines about 100 km east of Rome carries out solar neutrino physics, a search for proton decay, and a range of cosmic-ray experiments. A high-intensity electrostatic accelerator for nuclear astrophysics is a new addition to the Gran Sasso Laboratory. In the United States, in addition to Homestake, a large underground detector at the Sudan mine in Minnesota searches for proton decay and studies energetic neutrinos produced high in the atmosphere. There are new proposals from nuclear and particle physicists to construct even more massive detectors to study energetic neutrinos created in exotic galactic or even extra-galactic processes.

The next generation of solar neutrino detectors include BOREXINO, planned for the Gran Sasso and indicated in Figure 7.9. The detector is based on liquid scintillation spectroscopy, a standard method for detecting particles in nuclear and high-energy physics, and aims at a precise measurement of the neutrino flux from the ^{7}Be decay in the Sun. But the size of the detector and its requirement for exceptionally low backgrounds are not at all standard. A prototype detector, an order of magnitude smaller, has recently succeeded in the development of the necessary new, high-purity, low-background materials and techniques. A detector under the ice at the South Pole and an underwater detector project in Russia are pioneering new techniques. Because of the low background and unique sensitivity, experimental activities at underground laboratories address diverse and important questions, representing an exciting frontier of nuclear physics, a frontier that will likely grow in the next decade.

These are examples of some of the important instrumentation in use with nuclear physics experiments today. The scope of this report does not allow a comprehensive accounting of all developments under way or in use. Development of new schemes and novel concepts is continuously occurring. These

FIGURE 7.9 View of the prototype detector developed for the BOREXINO solar neutrino experiment at the Gran Sasso underground laboratory in Italy. The central 2-meter-diameter nylon sphere holds 4.5 tons of liquid scintillator and is surrounded by 1,000 tons of high-purity water. (Courtesy BOREXINO group.)

developments, as for new accelerator technologies, provide a unique and challenging training ground for the future scientists in nuclear physics and related sciences, but also for the technical workforce of the nation in general (see Box 7.1). Constantly pushing these frontiers, which have no technical peers, has generated and will generate the technical breakthroughs that are essential for progress and world leadership in the sciences, as well as the societal applications that many of these technologies achieve.

COMPUTERS IN NUCLEAR PHYSICS

Computers are an essential tool for research in nuclear physics. They are used for high-speed computing and for storage and retrieval of experimental data and other information. The needs of the field far exceed the capabilities of the present state of the art in computers and computational methods. Thus parts of the field are always exploring and pushing the limits of our computational abilities. They promote the development of new computational methods and new special purpose computers. The further applications of these methods have impacts in many other spheres. Some of the computational challenges facing nuclear physicists are described below.

Relativistic Heavy-Ion Data Storage and Retrieval

The detectors under construction at the RHIC facility in Brookhaven will yield vast amounts of information because up to 10,000 particles can be created in a single collision of gold ions. Since millions of collisions must be recorded from the finely divided, highly granular detectors, the total accumulated data volume will be close to one million gigabytes per year. These data sets are ten times larger than from previous experiments in either nuclear or particle physics, and their storage will require large-scale, high-bandwidth disk farms, tape robots with high-speed tape drives, and massive tape vaults. Furthermore, an input-output capacity to record the data as the collisions occur, and throughput of data at the analysis stage, must also be an order of magnitude faster than the present state of the art. All this has to be seamlessly tied together, allowing physicists access to data through simple database queries. A special Computational Grand Challenge project on data access is under way to provide software to do this in collaboration with computing facilities at RHIC, SLAC, and CERN. It will develop more efficient techniques of data mining based on pattern recognition techniques that allow users to locate the particular type of events needed from the large body of stored data. Applications of these techniques beyond the physics community will likely be broad, as was the case with the World Wide Web.

Quantum Monte Carlo Simulations of Nuclei

Detailed predictions of the microscopic and global structures of nuclei, and of nuclear reaction rates are challenging because of the quantal nature of nuclei and the complexities of nuclear forces. Many new developments in computational physics, including some of the key developments in the quantum Monte Carlo methods now used in several branches of the physical sciences, were made to address these problems in nuclear physics.

Quantum Monte Carlo techniques, developed in the past few years, can predict the structures and simple reaction rates of light nuclei having up to 8 nucleons, using present supercomputers operating at about 10 gigaflops, i.e., performing 10 billion operations per second. These predictions are being used for the studies of nuclear forces, nuclear structure, and solar and primordial nucleosynthesis. The next goal is to address structure and reactions of the more complex carbon, nitrogen, and oxygen nuclei, having up to 16 nucleons in an ab initio fashion, starting with the basic interaction between nucleons, with anticipated future computers operating at speeds of several trillion operations per second.

Computer Simulations of Supernovae

Core-collapse supernovae are spectacular stellar explosions that mark the end of a star's life after millions to billions of years of stellar evolution, disrupting it almost entirely. They occupy a special place in astrophysics and cosmic hierarchy for many reasons. They are the most energetic explosions in the cosmos, releasing energy in the form of neutrinos at the staggering rate of 10^{45} watts! They are also responsible for disseminating and producing most of the heavy nuclei in the universe, without which life as we know it would not be possible.

Current supernova modeling revolves around the idea that the shock wave, which will ultimately disrupt the star and produce the supernova, is partly fueled by neutrino heating of the matter behind it. These neutrinos are created in the core of the hot proto-neutron star left behind by the shock, and the most challenging task is to calculate the neutrino transport taking into account the three-dimensional hydrodynamic motions of the stellar matter. This presents a fundamental scientific problem that can be solved by (and only by) stretching computational resources to the limit.

Lattice Quantum Chromodynamics

For the understanding of the quark-gluon structure of the protons, neutrons, and other hadron building blocks of matter, and that of the phases and behavior of hadronic matter under extreme conditions, quantitative answers are within reach.

The only known way to calculate quark-gluon bound states and dynamics is by numerical evaluation on a discrete space-time lattice of large dimension. This technique was originally developed by particle physicists and is now being applied widely to problems in both particle and nuclear physics. One challenging problem of particular interest in nuclear physics is the behavior of bulk matter at the very high temperatures characteristic of RHIC collisions.

Such lattice-QCD calculations pose an extraordinarily demanding computational physics problem. In recent years these demands have lead to constructing a series of the fastest civilian computers in the world, with theoretical physicists playing a key role. In addition, physicists have pioneered a number of algorithmic developments to obtain great improvements in the accuracy of the calculations. Progress has been impressive, but a great deal more needs to be, and will be, accomplished in the coming decades.

OUTLOOK

Progress in nuclear physics is intimately linked to the continuous development of sophisticated and high performance instrumentation, especially particle accelerators, particle detectors, and data acquisition and computer systems. This development entails a broad range of modern technologies.

The design, construction, and operation of the sophisticated devices used in nuclear research involve students at all stages. This direct participation enables the students to acquire the skills, experience, and vision to develop the next generation of instrumentation for use in nuclear science and applied research, and for industrial applications. Although they are motivated by basic research, accelerators and detector techniques used by nuclear scientists have a broad range of applications outside of nuclear physics, especially in medicine and materials science.

The education of young scientists, engineers and technicians is essential for the technological advancement of this instrumentation, for the progressive evolution of the field, and for meeting society's technological needs.

8

Nuclear Physics and Society

INTRODUCTION

Society supports fundamental research in the expectation of benefits that support national priorities. These benefits take many forms. Satisfying natural human curiosity about the workings of nature is one, and this is the principal motivation for most researchers. Their search for new knowledge often stimulates advances in the limits of technology. It leads to instrumentation and theoretical concepts that address problems of societal concern, and to advances in other areas of science. The concepts and techniques of nuclear physics have had exceptional impact in this regard.

An equally important aspect is the contribution nuclear physics makes to the education of the technically sophisticated workforce that is essential for the nation's present and future economic well-being. Graduate education in nuclear physics provides broad training, involving experimental and conceptual techniques from a broad range of science and technology. As a result, nuclear physicists contribute in many areas of our society, frequently well beyond their original training in nuclear physics. Nuclear physics laboratories also provide an infrastructure for the hands-on education of younger students, involving undergraduates in research and exposing secondary school teachers and their students to the subatomic world and to scientific research.

The direct applications of nuclear physics have a major overlap with the priorities of the nation: improvements in human health, the environment, the efficiency of industrial processes, energy production, the exploration of space, and national security. Beyond these direct applications is the general benefit that arises from pressing forward the frontiers of high-technology development.

Some of the most pervasive applications of nuclear physics are in medicine. Medical imaging techniques now widely used, such as positron emission tomography (PET) and nuclear magnetic resonance imaging (MRI), provide information in three dimensions about the structure and biochemical activity of the human interior. Radioactive isotopes produced by accelerators and reactors are routinely used in medical diagnostic procedures, in treatment, and in medical research. Cancer radiation therapy mainly uses electron accelerators and radioactive sources. Treatment with protons, neutrons, and heavier ions is becoming more widespread and shows great promise for improved selectivity and effectiveness.

Many applications to environmental problems take advantage of the exceptional sensitivity of nuclear techniques such as accelerator mass spectroscopy to obtain information not available by other means. One can determine oceanic circulation patterns, the rate of carbon dioxide exchange between the atmosphere and the land and oceans, and the historic climate record. All of these have major implications for an understanding of climatic change. Studies of groundwater resources and their recharge rates, and of the origin of atmospheric pollutants, also provide unique information.

The assortment of industrial applications reflects the great variety of industrial processes. One common theme is the use of nuclear techniques and accelerators to determine the composition and properties of materials, their structural integrity after manufacture, and their wear in use. Another is the development of techniques for the modification of materials through accelerator ion-implantation, as in the doping of microelectronic circuits, or the introduction of defects to increase the current-carrying properties of high-temperature superconductors.

Safety and national security are areas with broad applications of nuclear techniques. Their use in detection of explosives and weapons has occupied increasing attention as a barrier to terrorism. Diagnostic procedures based on nuclear physics techniques will play a major role in noninvasive monitoring of chemical weapons and in controlling the distribution of enriched uranium and plutonium from dismantled nuclear weapons. Such procedures will also be important in the stewardship of the remaining nuclear stockpile. Intense beams from accelerators may in the future serve a joint role in production of the tritium required to maintain the required stockpile of nuclear weapons and in disposal of radioactive wastes.

Nuclear physics continues to have a profound impact on the production of energy: nuclear fission reactors produce about 19 percent of U.S. electricity (17 percent worldwide), and they provide an option for reducing use of finite hydrocarbon fuels and hence the emission of carbon dioxide into the atmosphere.

A few examples—of successes, of programs in early stages of development, and of some others with a good chance to become important in the future—are given here.

HUMAN HEALTH

Technologies emerging from nuclear research have an important impact on human health and have resulted in a new field, nuclear medicine. In the United States, 1,600 radiation oncology departments operate 2,100 linear accelerators. Nuclear diagnostic medicine generates approximately $10 billion in business annually, radiation therapy using linear electron accelerators about the same, and instrumentation about $3 billion. Over 10 million diagnostic medical procedures and 100 million laboratory tests using radioisotopes are performed annually in the United States. Three areas of particular medical significance are cancer radiation therapy, diagnostic imaging, and trace-isotope analysis.

Radiation Therapy for Cancer

Over a million new patients develop serious forms of cancer every year, and about half of them receive some form of radiation therapy. Traditional radiation treatments use streams of x rays or nuclear gamma rays. These high-energy photons deposit most of their energy where they enter the body. Thus, for a single exposure, healthy tissue unavoidably receives a higher dose than the cancer. The damage to healthy tissue can be ameliorated by irradiating the tumor from many different directions, all intersecting at the site of the tumor. Teams of radiologists, physicists (many with training in nuclear physics), and computer programmers design three-dimensional treatment plans (conformal therapy) that maximize dose-deposition in the tumor while minimizing the exposure of healthy tissue.

Recent developments by nuclear scientists and radiologists that use protons, neutrons, and heavy ions for radiation therapy promise to reduce the problems inherent in treatment with photons. As new accelerators designed explicitly for cancer treatment come into wider usage over the next decade, it is likely that there will be significant improvements in radiation treatment, resulting in more cures and fewer side effects. These advances will use techniques, knowledge, and accelerators that stem from nuclear physics.

Cancer Therapy with Protons

The use of protons for radiation therapy has the advantages that protons deposit more of their energy where they stop, not where they enter the body, and that their depth of penetration can be precisely controlled so that they stop within the tumor. This allows radiologists to increase the radiation dose to the tumor while reducing the dose to healthy tissues.

Over 20,000 patients have been treated with protons, mostly at accelerators originally built for physics research. Now, physicists are designing accelerators optimized for cancer therapy; one has been in operation since 1990 at Loma Linda Hospital near Los Angeles, and many others are in various stages of planning and construction, both in the United States and overseas.

Cancer Therapy with Neutrons and Heavy Ions

Research is continuing with other forms of radiation therapy that use neutrons and heavy ions. Neutrons produce a high linear energy transfer (LET); i.e., the density of broken chemical bonds in the cell is high. High-LET radiation overcomes a cancer cell's resistance to radiation damage more effectively than low-LET photon, electron, or proton radiation. Thus neutrons appear to be more biologically effective in killing cancers than are many other forms of radiation, especially in oxygen-poor cells. After three decades of clinical experience, it appears that some 10 to 15 percent of patients referred to radiotherapy would benefit from neutron therapy for cancers such as salivary gland tumors, some head and neck tumors, advanced tumors of the prostate, and melanomas.

A recent example of an optimized neutron facility is the superconducting neutron-therapy cyclotron designed and constructed by the National Superconducting Cyclotron Laboratory at Michigan State University. This cyclotron is in operation at Detroit's Harper Hospital, where neutron therapy is part of new cancer treatment protocols that have already shown highly promising results for tumors otherwise difficult to treat.

Beams of heavy ions, such as carbon or neon, with energies of 400-800 MeV per nucleon, are nearly ideal dose delivery vehicles for radiation therapy. They produce high LETs, which may offer the advantage of selectively destroying cancer cells (as compared to normal cells), and a sharply defined dose profile. Limited studies with carbon and neon beams were conducted at the Bevalac accelerator at Berkeley, but the studies were insufficient to establish a clear clinical advantage over proton therapy. Following pioneering U.S. studies, clinical research to assess the effectiveness of heavy ion radiation therapy using new accelerators and/or new techniques is now being pursued in Japan and at the GSI laboratory in Germany, but not in the United States.

Diagnostic Imaging

Diagnostic imaging technology started a century ago when Roentgen, the discoverer of x rays, immediately applied the penetrating power of these high-energy photons to make images of the interior of the human body, thereby revolutionizing diagnostic medicine. New imaging techniques have continued to have revolutionary impacts by providing improved, more sophisticated ways to see inside the body without surgery. Fundamental discoveries in physics have given us x rays, computerized axial tomography (CAT), nuclear magnetic resonance imaging (MRI), single photon emission computerized tomography (SPECT), and positron emission tomography (PET). Some of these—CAT scans, MRI, and, increasingly, SPECT—are now standard diagnostic tools of comparable importance with basic x rays, and the needed instruments are provided by commercial manufacturers. The practitioners of advanced PET techniques are often nuclear physicists, who continue to develop more powerful instruments and techniques, and work with physicians to apply the techniques in the medical environment.

SPECT and PET Imaging

The SPECT imaging technique uses drugs containing small amounts of short-lived radioactive isotopes, mainly single photon emitters. The emitted photons are viewed by large detector arrays, which are moved around the patient to obtain a complete picture of the drug's concentration in the body. If the drug accumulates only in particular sites, such as cancer metastases, the images show the location of these metastases (Figure 8.1).

In contrast, the PET imaging technique uses drugs containing small amounts of short-lived radioactive isotopes that emit positrons. When a positron encounters an electron in the patient's body, the two particles annihilate and emit a pair of photons, which move in opposite directions and strike radiation detectors arranged in a circle around the patient. The line connecting the two detectors that were struck passes through the point where the annihilation took place; by using information from all detectors, one can pinpoint the location of the nuclear decay. Since the annihilation takes place near the drug molecule that contained the positron-emitting nuclide, PET devices can image metabolic activity within the human brain for neurological and psychiatric evaluations, or the whole body for detecting cancer, or the metabolism in the heart and other organs. One can study the body in near-equilibrium by administering the positron-emitting nuclides slowly with time, or study the body's dynamic response by administering the positron-emitting nuclides over a short time interval and then observing their spread through the body with time. Physicists are currently developing ultra-fast PETs that could one day be used for online dose verification in cancer radiation therapy, allowing much more accurate dose administration than is now possible.

Nuclear Magnetic Resonance Imaging

An important activity that helped unlock the mystery of nuclear structure was systematic study of nuclear magnetism. In its simplest manifestation, the nucleus behaves like a tiny bar magnet. Developing precision methods for measuring the strength of the nuclear dipole magnet was an early goal of nuclear physics. This basic research challenge was met with elegant experimental tech-

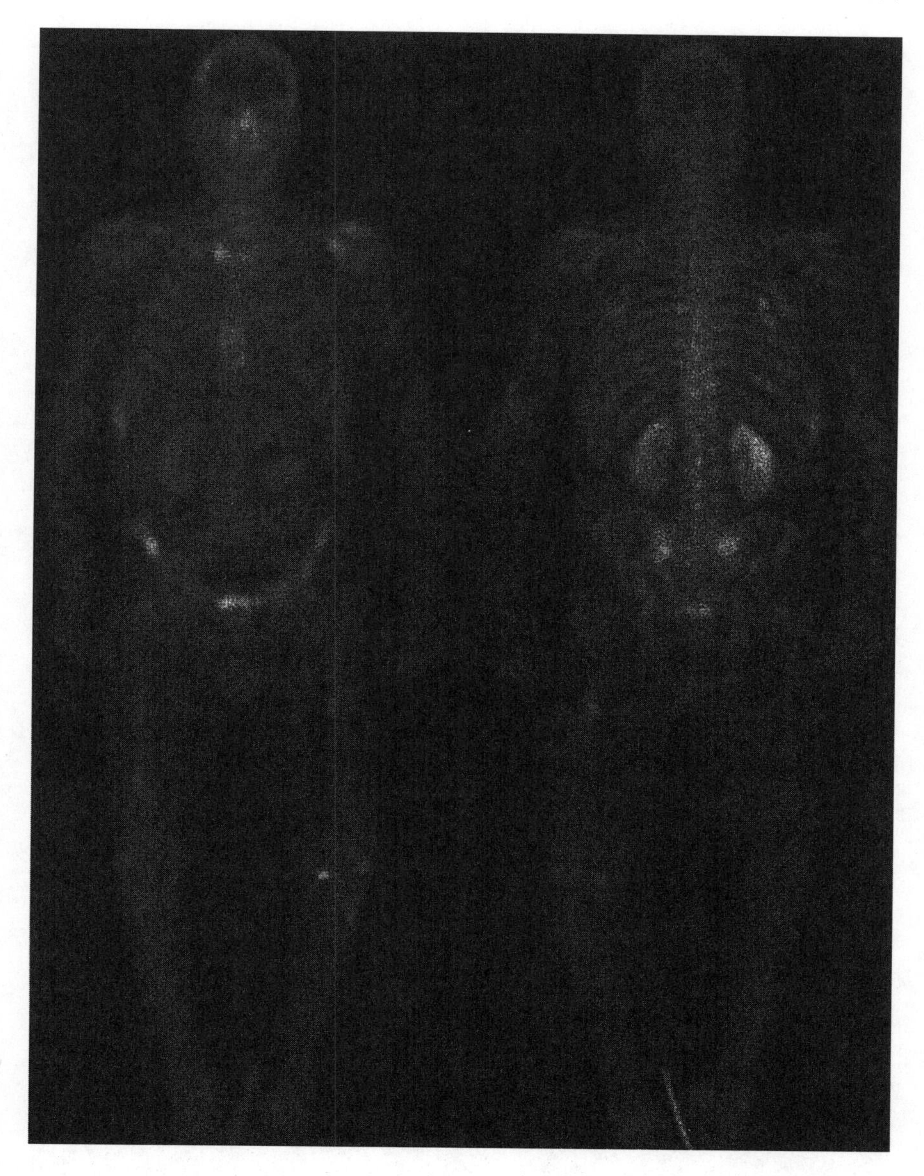

FIGURE 8.1 SPECT image of a man (front and back), showing increased uptake of a radioactive barium tracer as bright spots. This is interpreted to mean that a cancer has spread to the skeleton. (Courtesy of Mallinckrodt Institute of Radiology, Washington University Medical Center.)

niques that have had enormous impact on diverse areas of science and technology. Even in the earliest experiments, it was found that measurements of nuclear magnetism were affected by the environment. It was necessary for the nuclear experimentalists to understand the environmental effects in order to extract the nuclear information that they desired. The modern technique of magnetic resonance imaging developed from this area of basic nuclear physics research. MRI is used to detect these environmental influences and to obtain images of the body's interior from them, providing a remarkable medical diagnostic tool.

New opportunities are being provided by nuclei that have their magnetism (spins) oriented in a particular direction. As discussed in Chapters 2 and 3 of this report, polarized protons and nuclei provide greatly improved sensitivity in studying the structure of matter. Motivated by such research opportunities, nuclear physicists, working with atomic physicists, have recently developed highly polarized nuclear targets using finely tuned lasers that generate polarized light of precise wavelengths. They have then used these targets at a variety of accelerator facilities in the United States and abroad (e.g., MIT-Bates, IUCF, SLAC, DESY in Hamburg, and shortly, TJNAF) to probe the structure of nucleons and nuclei. Some of these nuclei, the gases ^{3}He and ^{129}Xe, have the magnetic properties needed for MRI and the atomic structure needed to retain their polarization for hours at a time. They can be introduced into lungs, allowing MRI studies of lung function. In recent experiments, a group of atomic and nuclear physicists and medical researchers used this technique with ^{3}He and ^{129}Xe to obtain MRI images of human lungs (Figure 8.2). Because of the strong signal provided by the polarized nuclei in the gas atoms, the MRI scans are short and can be synchro-

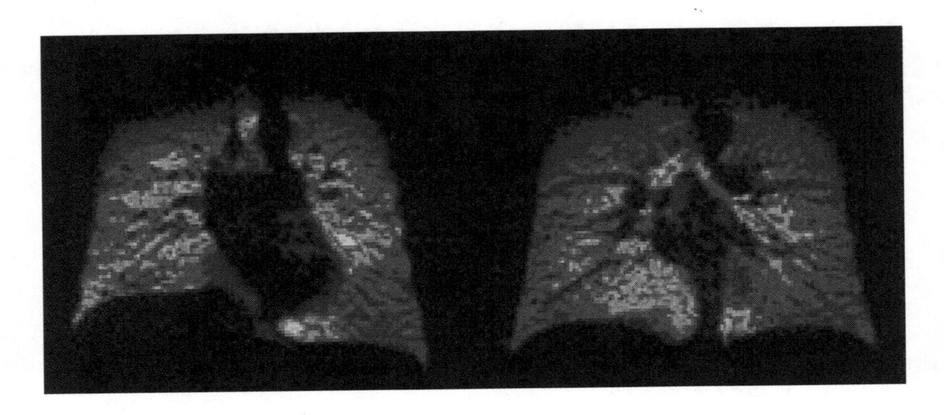

FIGURE 8.2 Nuclear magnetic resonance (MRI) images of a human lung, made possible by inhaling polarized ^{3}He gas. Because roughly 10 percent of the magnetic moments of the gas are lined up in one direction, a high sensitivity is achieved. This makes it possible to view the largely empty space of the lung. (Image courtesy of Duke University and Magnetic Imaging Technologies, Inc.)

nized with breathing. MRI with polarized nuclei provides the only imaging method now available for examining lung function. ^{129}Xe provides another capability; it dissolves well into blood and may allow the study of biochemical detail throughout the body.

Trace-Isotope Analysis

Radioactive nuclear isotopes produced by accelerators or nuclear reactors are used in many areas of biological and biomedical research. These isotopes have chemical properties essentially identical to their stable counterparts, but they decay and emit characteristic radiation that is readily detected. By inserting such radioisotopes as ^{14}C and tritium, it is possible to turn molecules into tiny transmitters without perturbing their natural biochemical properties. The signals from these transmitters (their unique radioactive decays) provide information on how molecules move through the body, what types of cells contain receptors, and what kinds of compounds bind to these receptors. Radioisotopes help researchers to develop diagnostic procedures and to help create new pharmaceutical treatments for diseases, including cancer, AIDS, and Alzheimer's disease. They are also used directly to treat disease. Radioactive tracers are indispensable tools for the new forensic technique of DNA fingerprinting, as well as for the Human Genome Project, which seeks to unravel the human genetic code.

Accelerator Mass Spectrometry

Accelerator mass spectrometry (AMS) was developed by nuclear scientists building on the experimental technology of nuclear physics. It is used in a number of areas of research, medical research among them. The technique uses nuclear physics accelerators to make possible new uses of isotopes in the health sciences, for applications where the common techniques are inadequate. In AMS, atoms from a minute sample are ionized and accelerated to a sufficiently high energy that one can detect and identify individual atoms, using nuclear techniques. One thereby measures the concentration of a given tracer without having to wait for its decay. When the time available for observation in the laboratory is much shorter than the half-life of the isotope, AMS has a much higher sensitivity for long-lived isotopes than does decay counting. Only very small quantities of tracer material are required, greatly reducing exposure to radioactivity.

This technique has had many applications in other fields of science, for example in geology, oceanography, archeology, and climate studies. Although most AMS work uses dedicated low-energy accelerators, advanced facilities used for basic nuclear research often extend the technical edge in terms of sensitivity and mass range. For example, an isotope of krypton (^{81}Kr) produced by cosmic rays has a lifetime of 210,000 years and will be useful for groundwater dating. It was necessary to use a large heavy-ion accelerator at MSU to produce Kr at a sufficiently high energy so it could be identified at its natural level of about 5 atoms in 10^{13} atoms of stable krypton.

In biological applications of AMS, ^{14}C can be used as a tracer with one million times the sensitivity of conventional scintillation counting. With this advantage, one can determine the uptake of hazardous chemicals through the skin or measure the damage to DNA by carcinogens or mutagens at actual exposure levels; environmental hazards or safety risks of pharmaceuticals can be evaluated without requiring (unreliable) extrapolations from unrealistically high doses. For example, the amount of benzene from a single cigarette has been traced in vivo to the exact proteins in the bone marrow of a mouse that this toxin affects. One can also use rare, but naturally occurring, isotopes to assess disease states. For example, the long-term progression of bone loss from osteoporosis can be studied at very low radiation exposure to the human subject by detecting minute amounts of the long-lived rare isotope ^{41}Ca. And it is now possible to optimize the dose of a drug for an individual patient by measuring the fraction of an isotopic tag in excreted metabolites.

ENVIRONMENTAL APPLICATIONS

Accelerator mass spectrometry is also an important tool for environmental measurements. Measurements that would otherwise be difficult or impossible are made routine by its great sensitivity. One important example is the use of long-lived radioactive nuclei to obtain information about past and present climate.

Cosmic rays—mainly high-energy protons—from elsewhere in the galaxy continually bombard Earth's atmosphere and surface, producing long-lived radioactive nuclei. These cosmogenic nuclei can be used to provide information that cannot be obtained by other means. The best-known application involves the isotope of carbon with mass 14. Because carbon in organic objects is not replenished from the atmosphere once the animal or plant dies, the ^{14}C present decays with a 5,700 year half-life, and the amount remaining provides a measure of the object's age. Other cosmogenic nuclei can be used in a similar manner to determine how long material that contains them has been shielded from cosmic rays and from the atmosphere. The concentration of the long-lived isotope ^{81}Kr in an aquifer of the Great Artesian Basin in Australia will be measured and used to determine how long its water has remained uncontaminated with younger groundwater. Accelerator mass spectrometry has greatly advanced the science of dating; its sensitivity allows the use of extremely small samples. A previous attempt to detect ^{10}Be and ^{26}Al in ice required 100 tons of ice—with AMS, 1 kilogram is sufficient.

Some of the most important applications of cosmogenic nuclei are studies of large-scale environmental phenomena. The amount of ^{10}Be in ice cores has been measured by AMS and is found to be correlated with solar activity. This correlation may make it possible to extend studies of solar activity backward 10,000 years in time, compared to the 400-year record currently available. A determination of whether solar variation could be partly responsible for climate variation may then be possible.

Other extensive measurements are devoted to understanding the nature of oceanic circulation. The pattern of these ocean currents is shown in Figure 8.3. Because they transport large amounts of water and heat, changes in the circulation pattern can have a major influence on climate. For example, if northward-flowing currents in the Atlantic were suddenly to cease, the temperature in northern Europe would decrease by 5 to 10 °C. An accurate description of these currents is of great interest, because of a concern that the driving force of increasing greenhouse gases could initiate such a change, causing sudden stepwise changes in climate or a shift to a climate regime less stable than the present.

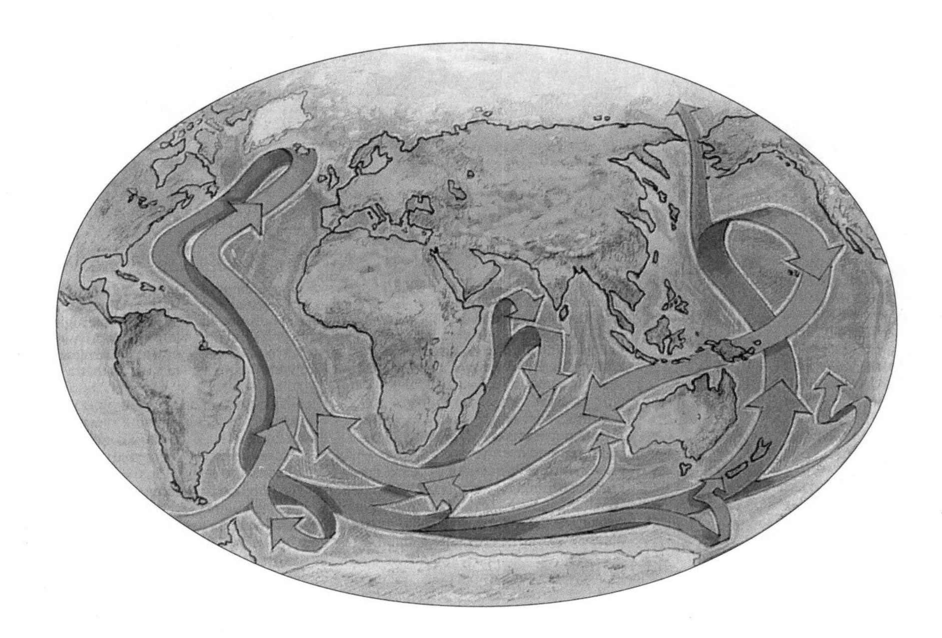

FIGURE 8.3 Global currents carry large amounts of heat. Cold, salty water (dark shading) initially formed in the North Atlantic sinks into the depths of the ocean and flows south. The warm water (light shading) that flows northward to replace it is responsible for the present warm climate of northern Europe. It seems possible that these currents may stop, leading to rapid climatic changes. (From "Chaotic Climate," by W.S. Broecker, *Scientific American*, Nov. 1995, page 62.)

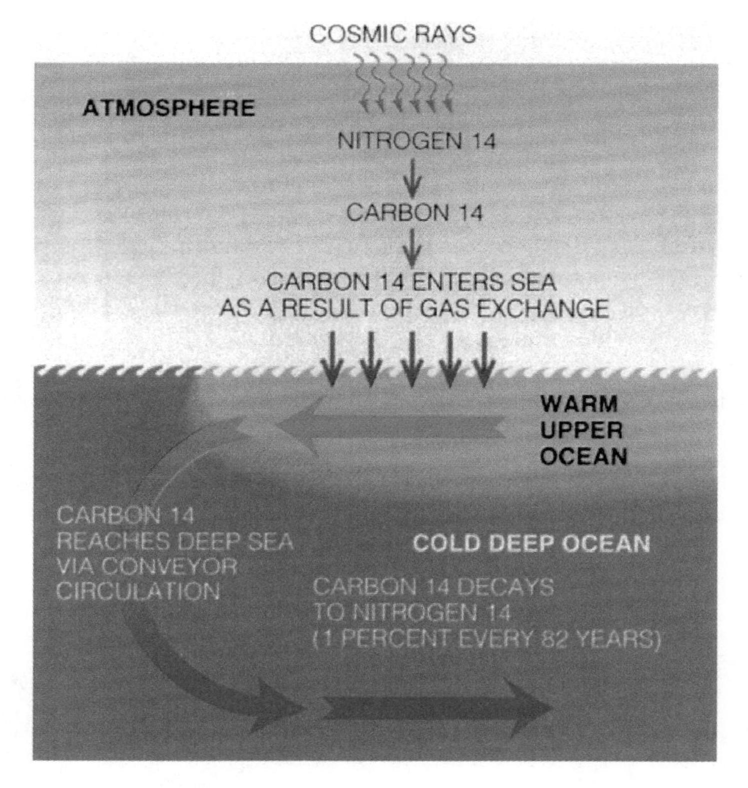

FIGURE 8.4 As cold, salty water sinks to great depths, it carries radioactive ^{14}C out of the atmosphere and into the abyss, where it slowly decays. Radiocarbon dating is used to measure the state of the oceanic current system. (Courtesy Barry Ross.)

Measured concentrations of oxygen isotopes in Greenland ice cores show that large changes were common near the end of the last ice age. Dating of organic glacial remains in New Zealand using ^{14}C indicates that these large changes were global in nature.

As shown in Figure 8.4, ^{14}C enters the ocean and is carried into the abyss by descending currents. Precise measurements of ^{14}C yield the mixing times for deep water and information on the stability of the oceanic circulation pattern.

A different application is to measure the concentrations of long-lived isotopes produced by human activities and deduce the exposure to more dangerous short-lived isotopes of the same element long after they have decayed. This technique made it possible to reconstruct deposition patterns and thyroid doses from radioactive ^{131}I a decade after the Chernobyl accident. Although ^{131}I deposition in the thyroid had been measured for more than 100,000 people soon after the accident, its half-life is so short (8 days) that it was impossible to repeat or extend these measurements. Since the half-life of ^{129}I is 16 million years, it can still be detected by AMS, and together with the known ratio of ^{129}I and ^{131}I emitted by the reactor, used to predict thyroid exposure to ^{131}I. So far, AMS is the only way to do this with the necessary accuracy.

IMPACT ON INDUSTRY

Techniques derived from experimental nuclear physics and its accelerators and detection devices are today pervasive in technical and industrial applications. They are used for diagnostic and testing purposes, for material modifications, and directly for production processes. In addition, nuclear physics research provides vital input data for applications and engineering designs.

Nuclear Analysis and Testing

Certain techniques of nuclear physics—accelerators for producing a wide variety of particle beams, and methods for detecting and characterizing a broad range of nuclear products—are the basis for many applications. Some of these are discussed elsewhere in this chapter. Here, we concentrate on a small sample of important applications in industry.

Testing with Particle Beams

When a particle scatters backward from another particle, its energy depends on the mass of the target particle. A technique known as Rutherford back scattering, routinely used in industry, takes advantage of this property to determine the elemental composition of a material as well as the depth distribution of various elements in the bulk material. Rutherford's scattering technique was a monumental scientific accomplishment. The scattering experiment was invented to answer a simple but important basic research question: How is electric charge distributed within the atom? In answering the puzzle, Rutherford discovered the atomic nucleus and pointed out the power of a scattering technique that has led to important technological advances in many areas of science and technology. Recently, Rutherford scattering techniques have enabled important practical advances in the semiconductor industry where they are used to characterize the minute but important impurities in semiconducting materials. Materials also emit x rays and gamma rays that are characteristic of the material when bombarded by low-energy protons. The x-rays (proton-induced x-ray emission or PIXE technique) or gamma rays (proton-induced gamma-ray emission or PIGE technique) are used to detect the presence of specific elements with high sensitivity and spatial resolution. These techniques have an enormous range of applications;

FIGURE 8.5 The alpha-proton-x-ray spectrometer points toward us at the lower center in this picture of the Mars Rover. It consists of an alpha-particle source and detectors to observe the back-scattered alpha particles (Rutherford back scattering) as well as the protons and gamma rays produced by alpha bombardment. Together, this information allows one to determine the composition of rocks on Mars. (Sojourner™, Mars Rover™, and spacecraft design and images copyright © 1996-97, California Institute of Technology. All rights reserved. Further reproduction prohibited. Reprinted by permission.)

recently an alpha-proton-x-ray analyzer installed in the Mars roving vehicle Sojourner analyzed the composition of martian rocks (Figure 8.5).

Testing with Neutron Beams

Following capture of a neutron, nuclei emit gamma rays that are characteristic of the nucleus. It is possible to produce copious beams of neutrons by using low-energy nuclear reactions, such as the deuteron plus triton reaction, bombard an unknown sample with the neutrons, and detect the presence of specific materials tagged by their characteristic gamma rays. The known dependence of total interaction probabilities on material provides another possible tag. Airport safety devices may use these techniques to check for the presence of nitrogen in otherwise undetectable plastic explosives. Neutron techniques have been especially refined for oil-well logging and are widely used for this purpose. The neutron generators must be compact, and the instruments must be able to withstand pressures as high as 2,000 times atmospheric pressure and temperatures up to 175 °C.

Materials Modification

Beams of high-energy particles and gamma rays have many applications in industry. For example, gamma rays are used for the sterilization of foodstuffs and to cure epoxies. Implantation of beams of heavy ions developed for nuclear physics research is routinely used in the production process for semiconductor devices. As these devices become smaller, they become susceptible to faults caused by the electrical charges produced by background ionizing radiation; accelerator beams are widely used to study sensitivity to these single event effects.

A recent application is the use of proton, neutron, and heavy-ion beams for the improvement of high-temperature superconductors. These superconductors, with transition temperatures as high as 130 K (–143 °C) at ambient pressure, lose all electrical resistance and become superconducting if cooled by liquid nitrogen. For most applications, such superconductors must be able to carry high currents with zero dissipation. In high-T_c superconductors, magnetic fields produced by such currents penetrate into the material and establish tiny magnetic flux tubes that are surrounded by circulating supercurrents. In the presence of an electric current, these flux tubes move, producing thermal losses, unless pinned to a defect in the material. Neutrons or heavy-ion bombardment can produce such pinning defects and thus greatly enhance the current-carrying capability of the high-T_c material. Heavy-ion beams with energies between 5 and 100 MeV per nucleon produce the strongest pinning centers, enhancing the current-carrying capability by several orders of magnitude. Experiments on flux pinning are performed today at many nuclear research centers.

U.S. Nuclear Data Program

As should be clear from the above discussion, nuclear phenomena and techniques have a broad range of applications. Another by-product of nuclear research is a variety of nuclear data, such as nuclear-level schemes and moments and nuclear interaction probabilities, that play a major role in applications. For example, the design of nuclear reactors requires a knowledge of the details of the interaction of neutrons with reactor fuels and materials, and of the decay properties of many of the radioactive nuclei that are formed. To provide this information, the DOE operates a Nuclear Data Program that, in the context of an international collaboration, is charged to collect, evaluate, and tabulate data useful for applications and for fundamental research.

ENERGY

Large amounts of energy can be released by splitting or fissioning heavy nuclei, such as uranium, or by fusing light nuclei, such as the isotopes of hydrogen. For example, a very large fission power plant producing a billion watts of electrical power will consume only a ton of uranium in a year. In the 1960s and 1970s, a large number of nuclear fission reactors were constructed to take advantage of this energy source; at present about 110 U.S. reactors are in operation. In 1996, nuclear energy provided about 19 percent of the nation's electric power production. That share has grown continuously, without construction of new plants, because the reliability of nuclear reactors has improved to 76 percent in 1996. Several states obtained more than 50 percent of their electric power from fission in 1996. Internationally, France obtains the highest fraction of its electricity, 77 percent, from nuclear reactors (Figure 8.6).

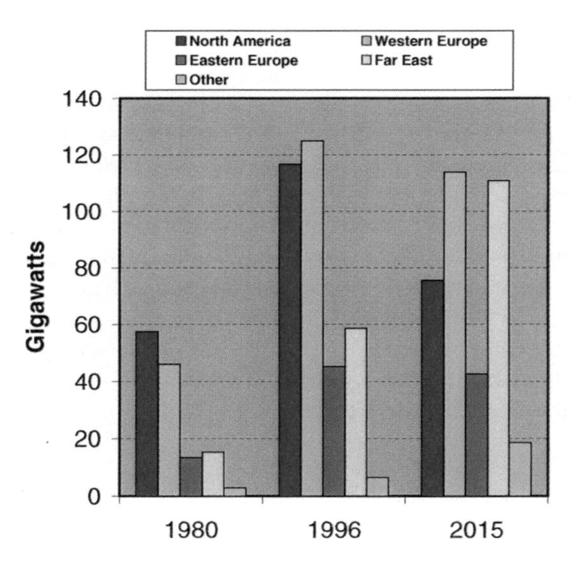

FIGURE 8.6 World nuclear capacity by region in 1980, 1996, and 2015. For a constant capacity, power output has increased in recent years because of increased operating efficiency. The expected decrease in capacity in the future is due to retirement of older reactors.

Since the 1970s, new construction of power reactors in the United States has come to a standstill. Concerns about operational safety and waste disposal have overshadowed the inherent advantages of nuclear power. These issues must be resolved before nuclear power can again be seriously considered on a large scale. Fission and fusion processes produce energy without producing greenhouse gases and components of acidic rain, such as sulfur—matters that are also of concern to society. Ultimately, this is an issue for public policy, and in the meantime it is important for the nation to preserve its options.

Several major new developments, described below, are concerned with improving reactor safety and the handling of waste; some of these will be tested during the next decade. Many of these developments draw on technologies and expertise that were developed as part of basic nuclear research programs.

Burning of Long-Lived Waste and Accelerator-Driven Reactors

One option for handling nuclear waste is to transmute the long-lived radioactive wastes from light-water reactors (the main reactor type) into shorter-lived isotopes that can be dealt with more easily. It has been proposed that intense high-energy proton beams could be used for this purpose; such transmutation machines have been studied at Los Alamos National Laboratory, at Brookhaven National Laboratory, and in Europe.

In one proposed process, proton beams would strike a high-power target and produce copious streams of energetic neutrons that would transmute the bulk of the radioactive waste material into short-lived nuclei. The goal is to reduce waste lifetimes to less than 100 years. Estimates indicate that with a further storage period of only 30 years, these products would have a level of radioactivity less than that of the original unused uranium fuel. The accelerator-waste combination would be operated at a subcritical level—it could not by itself sustain a chain reaction—so that no reactor-core meltdown accident could occur. In another proposed scheme, the PHENIX Project, uranium and most of the plutonium would be separated prior to proton irradiation and used again for reactor fuel. The most important long-lived components of the remaining waste are isotopes of neptunium, americium, curium, and iodine, some with half-lives of 10,000 years or more. According to one estimate, a machine operating at 3,600 million watts of thermal power could process these waste isotopes for 75 light-water reactors.

It has been suggested that this concept might be carried one step farther, and a particle beam might be used to produce additional neutrons directly in a nuclear-reactor-like core. Versions of this concept have been studied at Los Alamos and by a European group. The core would be sub-critical, and the accelerator beams would provide sufficient additional neutrons to run the reactor. Because the neutrons would have high energy, an abundant and less enriched element, such as natural thorium, could serve as the fuel. Such a sub-critical configuration is inherently safe, and the long-lived waste is transmuted to short-lived nuclei in the process. A 1996 design study concluded that a 1-GeV, 0.03 ampere beam requiring 60 million watts of input power would produce 675 million watts of electric power, amplifying the input power by about a factor of 10. The thorium fuel would not require enrichment, but it would need to be recharged every 5 years.

Inertial Confinement Fusion Reactors

Nuclear fusion provides an alternative approach to producing clean and abundant power. In this process light nuclei such as deuterium and tritium fuse into helium, a process by which the Sun and other stars produce their energy. The fuel supply for fusion, the deuterium in the oceans, is extremely large, and fusion produces no long-lived radioactivity.

Since the 1950s, most studies have used magnetic fields to confine the plasma of interacting nuclei. Another technology, inertial confinement fusion, was proposed more recently: capsules of frozen deuterium and tritium (D-T) would be spherically compressed to several thousand times their normal density by powerful beams of lasers, or of light or heavy ions. Much has been learned about capsule fabrication and compression using powerful laser beams, and the National Ignition Facility, authorized for construction at Lawrence Livermore National Laboratory, has the goal of using lasers to heat D-T capsules to ignition. In the longer run, intense heavy-ion beams may offer a more practical solution.

Nuclear laboratories have much experience with the production and use of heavy-ion beams; their expertise would be important in the development of such powerful heavy-ion machines.

NATIONAL SECURITY

Nuclear science plays a critical role in national security. It is a matter of national policy that the maintenance of an effective nuclear stockpile continues to be important with the end of nuclear testing. Related to this issue are concerns regarding the proliferation of nuclear weapons and the acquisition of such weapons by terrorist groups.

Stockpile Stewardship

With the end of the Cold War and cessation of nuclear weapons testing, there is a new emphasis on stewardship of the existing nuclear stockpile that is science based rather than test based. It involves intensive computer simulations to replace underground explosions; these simulations rely heavily upon detailed understanding of the relevant nuclear physics processes and parameters.

The simulations must model the explosion of the warhead and also follow the explosion products to determine where and how the energy of the explosion is deposited. As a result, many reaction probabilities—for charged particles, x rays, gamma rays, and neutrons—as well as the nature of the beta-decay products of fission fragments, are needed to fully simulate the performance of a nuclear weapon. Many of these reaction probabilities and decay properties are known from measurements, but others must be calculated from an understanding of how nuclei behave. This requires a deep understanding of nuclear structure and reactions.

Stockpile stewardship also has an important experimental component. Radiographic techniques that can image warheads without opening them and that can measure the dynamics of nonnuclear explosions will play an important role. Proton radiography is a sharp departure from the flash x-ray technology that has been the predominant radiographic tool in the past. It draws on the skills and techniques of nuclear physicists. The protons are created by an accelerator, transmitted using magnetic lenses, pass through the object under study, and are detected, all using techniques developed for studies in basic nuclear physics. Among the advantages high-energy protons bring to radiography are their penetrating power, well matched to the imaging of dense objects, and their sensitivity to both material density and composition. One can obtain multiple snapshots of dynamical processes such as chemical explosions, including stereoscopic views.

An accelerator at Los Alamos National Laboratory, built for research in nuclear physics (LAMPF), is carrying out a series of experiments to test this concept. A proton radiographic image of the shock front formed in a chemical explosion is shown in Figure 8.7.

This test demonstrates some of the basic concepts of proton radiography. A new experiment now under evaluation at Brookhaven National Laboratory will extend static measurements to higher energy, as required for radiography of thick objects. Eventually a proton accelerator in the 50 GeV range will be required.

Still another issue related to maintenance of the nuclear stockpile is the production of tritium, a heavy form of hydrogen. Tritium is an essential ingredient of thermonuclear weapons, but it decays with a half-life of 12 years and must be continuously replenished. One possibility for production of tritium is to use an accelerator to provide a 100-megawatt proton beam and drive a spallation source. Such a facility could provide the required time-averaged neutron flux. The safety, flexibility, cost, and logistics may prove more attractive than the alternative of constructing a large nuclear reactor. The LAMPF accelerator is currently being used to validate this concept.

Nonproliferation of Nuclear Weapons

The large accumulated quantity of nuclear material and secondary waste products poses a significant challenge. Such materials are used as fuel in nuclear reactors, but some can also be used in nuclear weapons. The Treaty on Non-Proliferation of Nuclear Weapons specifies that every "non-nuclear weapon state"

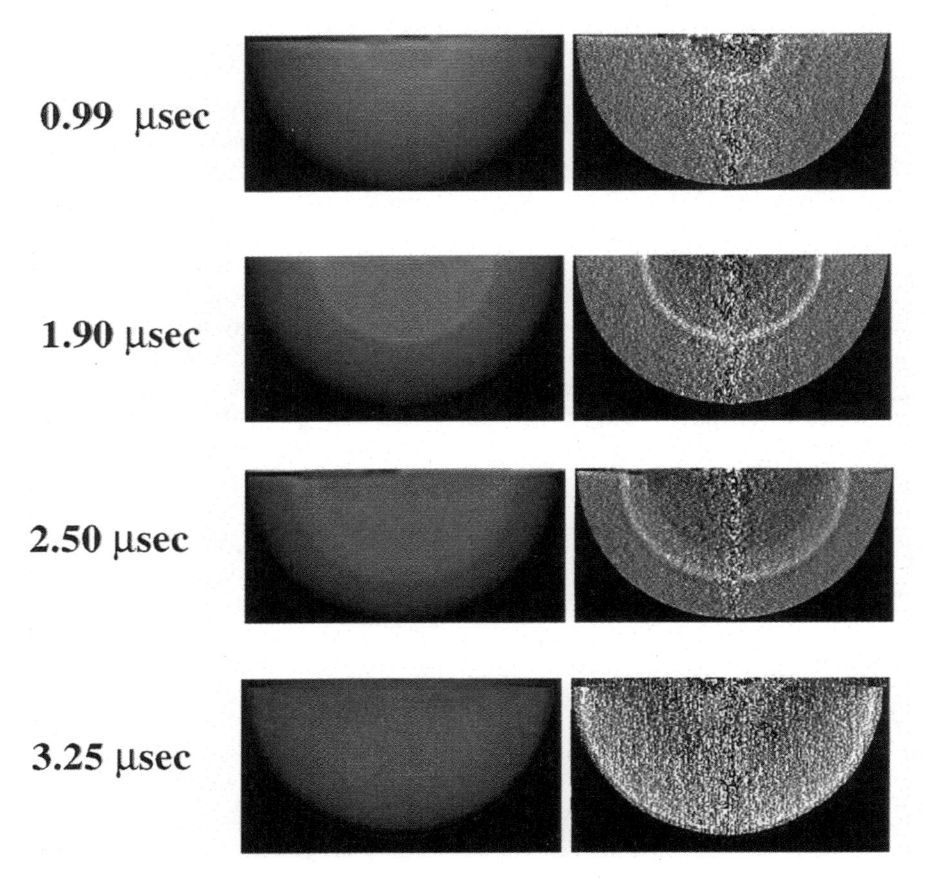

FIGURE 8.7 Proton radiograph of a chemical explosion, taken at the Los Alamos Neutron Science Center (LANSCE). In the sequence of exposures, at 0.99, 1.90, 2.50, and 3.5 microseconds, from top to bottom, one can see two different views of the progress of the shock wave formed in the explosion. (Courtesy Los Alamos Neutron Science Center.)

agrees to allow nondestructive analysis of their nuclear materials to prevent diversion from peaceful uses to weapons production. In those states with nuclear weapons, the material must be monitored and accounted for.

Techniques developed for nuclear physics research form the backbone of safeguards technologies. Gamma-ray detectors are used to assay the amount and isotopic composition of uranium or plutonium in a sample of material. Neutrons from fission are detected and independently determine the content of Pu in a sample. New techniques to enhance sensitivity include tomographic gamma scanning, analogous to medical imaging, and use of neutron sources composed of the element californium to measure the attenuation of the neutron flux by light elements in the object being assayed.

EDUCATION OF THE NATION'S TECHNICAL WORKFORCE

Education of the next generation of technically sophisticated citizens and the training of scientists who can contribute to society have high priority in the nuclear physics community. The unique assets of the research enterprise, particularly in nuclear physics, provide a superb infrastructure to address this priority.

Graduate Education in Nuclear Physics

Education of graduate students in nuclear physics is essential for the continuing health of the field, and for maintaining a technically sophisticated workforce with expertise in the many aspects of advanced technology and instrumentation that training in nuclear physics entails. Nuclear physics has a long tradition of producing broadly educated and flexible scientists. Their skills are readily applied to a wide range of the nation's technological problems, in business, industry, government, and medicine.

Students in nuclear experiment and theory have the opportunity to face and solve complex problems at the frontiers of knowledge. Their graduate training involves state-of-the-art instrumentation and knowledge from different fields, many outside of nuclear physics. As an experimentalist, a student commonly designs, builds, and tests hardware using advanced materials, vacuum technology, control technology, and complex electronics and semiconductor devices, and becomes expert at the electronic and computer technology in data acquisition and analysis. Both theoretical and experimental projects often involve the design and implementation of complex computer programs and some of the most advanced computer hardware. In many cases, students have responsibility for all aspects of their projects and thus acquire a broad range of skills in addition to the ability to attack a particular problem in depth. Students often interact with scientists from different institutions and different countries that may contribute detectors or other components to an experiment; they learn teamwork and management and communication skills in addition to acquiring new technical knowledge and expertise.

The field of nuclear physics continues to attract high-quality young people. At present, DOE and NSF are supporting about 650 graduate students in nuclear physics with perhaps one-third that number supported by other funds. This influx of talent advances the intellectual and technical forefront, and it makes nuclear physics an important source of technical manpower.

As an example, the rather complete records of three university-based nuclear physics accelerator laboratories show the career paths of nuclear physics Ph.D.s

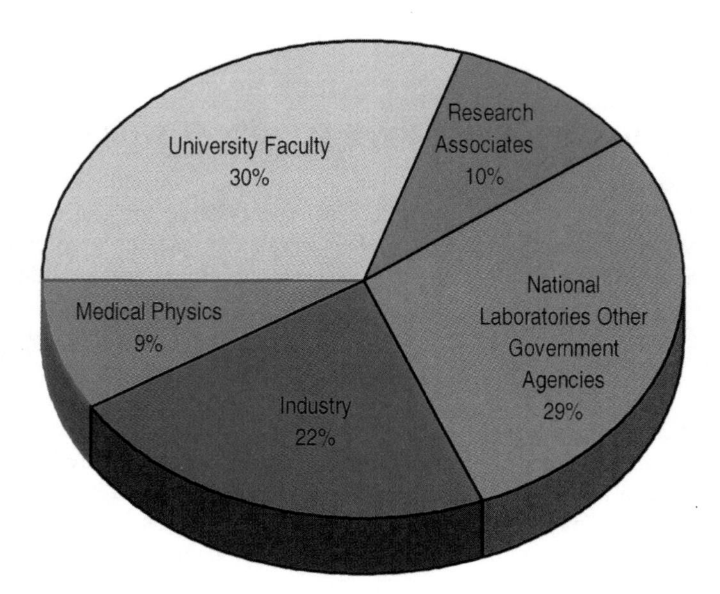

FIGURE 8.8 Employment of Ph.D. graduates in nuclear physics, from the long-term records of three DOE and NSF university laboratories: Indiana University Cyclotron Facility (IUCF), Michigan State University's National Superconducting Cyclotron Laboratory (MSU/NSCL), and the Triangle Universities Nuclear Laboratory (TUNL). They are probably similar to those of the field as a whole.

more quantitatively. These records, summarized in Figure 8.8, are probably typical of the field as a whole. The field of medical physics has long been a special and important application of nuclear physics and is displayed separately for this reason; most of these positions are in industry.

All the categories of permanent positions shown in Figure 8.8 include career opportunities whose scope is much broader than that of nuclear physics. This variety is not a recent phenomenon but has been characteristic of graduates in nuclear science for decades.

The nature of career prospects for young scientists is of considerable interest both within the scientific community and beyond. The decrease in funding for basic research and for the support of long-term research in industry has eroded a major sector of the traditional base of job opportunities for Ph.D. physicists in all fields. Statistics from studies by the American Institute of Physics indicate that the employment situation for physicists is now improving and that there has been a decrease in the number of first-year graduate students that will lead to a significant decrease in Ph.D. production. It appears that the entering generation of graduate students will have much-improved job opportunities. In the particular case of nuclear physics, the production of Ph.D.s has been stable, and it appears that their employment opportunities have been relatively constant over at least the past decade.

Graduate Student and Faculty Demographics

A number of surveys of people active in basic research in nuclear science have been carried out over the past 20 years, some based on questionnaires sent to essentially all funded researchers, and others on statistics provided by the funding agencies. Taken at face value these surveys indicate that the number of nuclear scientists may have increased since 1980—the largest increase is 18 percent in DOE-supported graduate students. But, because of their different methodologies, one cannot be certain that the different surveys are measuring the same quantities, and indeed, some of the results seem inconsistent.

A significant overall phenomenon is the drop in the number of senior physics majors and beginning graduate students in all fields of physics. There were 26 percent fewer first-year graduate students in 1996 than in the recent peak year, 1992. This will inevitably lead to a drop in the number of physics Ph.D.s. Such a drop clearly is a concern that transcends disciplinary boundaries; it is likely to result in increasing reliance on Ph.D.s from other countries.

Undergraduate Education

An undergraduate degree in physics provides an excellent preparation for many different career paths. Many universities have programs that prepare physics majors for the changing demands of the workplace. These include programs focusing on preparation for secondary education, business, and laboratory instrumentation, as well as interdisciplinary programs in chemistry, biology, and medicine and in environmental and public policy. One indication of the success of these programs in the past is shown in the results of a survey,[1] which showed that the median annual earnings (at mid-career) of holders of a physics bachelor's degree ranked among the highest-paying specialties and were 16 percent higher than the median for all fields.

The nuclear physics community has long provided undergraduates with experiences in active research laboratories, but in recent years this effort has had a sharper focus. Both NSF and DOE support such programs. Many students work with nuclear physicists through the NSF Research Experiences for Undergraduates (REU) program (see Figure 7.1.1). The DOE Science and Engineering

[1]Reported in "Earnings of College Graduates, 1993," Daniel E. Hecker, *Monthly Labor Review*, Bureau of Labor Statistics, U.S. Department of Labor, Vol. 118, No. 12, December 1995.

Research Semester and the Undergraduate Student Research Participation program involve students in nuclear physics projects. With the American Chemical Society, the DOE has developed an intensive summer school program that exposes students to research in nuclear power, waste disposal, nuclear nonproliferation, radiation safety, and nuclear medicine. Nuclear physics faculty members at undergraduate institutions often involve their students in nuclear research, especially at the user facilities, with NSF Research at Undergraduate Institutions (RUI) program support; about 9 percent of NSF-supported faculty are funded through the RUI program.

While these specific NSF and DOE programs provide opportunities for students who would otherwise not be able to participate in research, many more undergraduates are involved in nuclear physics research, directly supported by research grants. At a large number of universities, nuclear physics faculty involve undergraduates in their research projects, often as part of the research team, through senior theses and part-time or summer jobs. The synergism between research and teaching provides an early exposure to forefront research and state-of-the-art technology.

Earlier Education, Outreach, and Scientific Literacy

Given the rapid pace of technological advances, the future of the country and its economic welfare depend increasingly on the level of the technical and scientific sophistication of the population. The 1996 NSAC *Long-Range Plan* described the results of a survey on the involvement of nuclear physicists in undergraduate education, outreach, and scientific literacy. It is evident from the responses to the survey that many nuclear physicists are committing increasing amounts of their time and energy to these issues.

K-8 Education in Elementary and Middle Schools

Young children are fascinated by natural phenomena. Reinforcement of this fascination early in the educational process is a goal of many nuclear physicists. The broad variety of approaches includes programs for hands-on experience that have reached thousands of students; visiting minority professorships at research universities charged to interact with inner-city schools; and the Becoming Enthusiastic about Math and Science (BEAMS) program at TJNAF. The BEAMS program is a partnership with the local schools and the Commonwealth of Virginia; entire classes are brought to the facility for a full week of immersion in the scientific environment. To date, about 30,000 students have benefited from BEAMS and other programs at TJNAF.

Contact with Teachers and Students in High Schools

Students often perceive science and mathematics as formidable and as unpopular with their peers. This situation can be improved only if one addresses both the perception of science by young people and the quality of their science education. Nuclear physics laboratories and universities organize many programs for high school students and their teachers. Some of these efforts have been supported by federal funds or by the local institution, but many rely on the voluntary work of individual scientists. Examples include Saturday classes for in-service teachers; multiweek summer programs for students, sometimes involving teachers as well; lectures and demonstrations in schools, coupled with development of instructional material; and extension of computer facilities to high schools, so that students and teachers can access the Internet and the many innovative activities available there.

Activities Addressing Underrepresentation of Women and Minorities

It is unfortunate that large segments of our society are underrepresented in science and technology. The nuclear physics community has endeavored to encourage women and minority students to pursue careers in physics through individual volunteer efforts and specific programs supported by DOE and NSF. Many universities, colleges, and national laboratories bring female and minority students in middle schools for one day or longer visits to participate in hands-on science (especially physics) and to meet practicing scientists. Nuclear scientists have been active in such programs as the American Physical Society's Women in Physics project.

National and university laboratories have also committed resources to recruit students from historically black colleges and universities (HBCUs) and hispanic-serving institutions (HSIs) to participate in the laboratories' summer science research programs. These programs sometimes include support for HBCU faculty participation. For example, TJNAF's efforts have contributed to a significant growth in faculty hirings in HBCUs and HSIs. As a result, Hampton University has developed a new Ph.D. program and graduated about 20 undergraduates, all African American, one-third of whom has done research in nuclear physics. Mentoring programs for promising minority undergraduates have been instituted by national laboratories and universities.

It appears that the programs described above have had positive effects on an overall societal problem, but the issues remain.

OUTLOOK

The past direct contributions of nuclear physicists to problems facing the nation are substantial. This is surprising, given the direction of research in nuclear physics, which involves the most fundamental aspects of nature and is not directly focused on societal issues. If one examines the contributions outlined above, three threads running through them can explain this result. First, the techniques of nuclear physics are relevant to many of our national problems. Second, the broad training and team experience of many students in nuclear physics provide the background that allows them to confidently and fruitfully apply nuclear techniques in many settings. And third, the varied properties of nuclei, and their radiations, lend themselves to the remarkably broad range of specific applications discussed in this chapter.

It is appropriate to ask whether these contributions are likely to continue. The impact of basic research is hard to predict, but it can lead to profound and revolutionary developments, the case of nuclear fission being an outstanding example. Many of the items discussed above seem likely to have still greater importance in the future, and new applications will certainly arise from new technical developments in nuclear physics. One can anticipate continued growth in the role of nuclear physics in generating applications that contribute to society.

Appendix

Accelerator Facilities for Nuclear Physics in the United States

Tables A.1 and A.2 summarize nuclear physics accelerator facilities currently in operation or under construction in the United States. In addition to the accelerator parameters and performance characteristics, the tables list the primary areas of research that each one addresses. Figure A.1 gives an overview of facilities and their geographical location.

The major new facilities of the nation's nuclear physics program are the Continuous Electron Beam Accelerator Facility (CEBAF) at the Thomas Jefferson National Accelerator Facility, which recently came into operation, and the Relativistic Heavy Ion Collider (RHIC) at Brookhaven National Laboratory, which is scheduled to begin operation in 1999. CEBAF (shown in Figure 7.1) is a superconducting, recirculating linac designed to deliver continuous electron beams of up to 200 μA of current, polarized and unpolarized, simultaneously to three experimental areas. The design energy is 4 GeV, but operational experience with the superconducting cavities indicates that an energy of up to 6 GeV will be possible.

Nearing completion, RHIC (shown in Figure A.2) is the first colliding-beam facility specifically designed to accommodate the requirements of heavy-ion physics at relativistic energies. RHIC will provide heavy-ion collisions for a range of ion species up to gold, with beam energies of 30 to 100 GeV/nucleon for each of the colliding beams.

In addition to these two large nuclear physics facilities, six medium-size user facilities supported by DOE and NSF address different key aspects of nuclear physics, indicated in Figure A.2.

The Bates Linear Accelerator Center at MIT provides high-quality electron

TABLE A.1 National User Facilities

Facility	Beam Characteristics	
	Species	Energies
Thomas Jefferson National Accelerator Facility (VA) Continuous Electron Beam Accelerator Facility (CEBAF)	Electrons	1-6 GeV
Brookhaven National Laboratory (NY) Relativistic Heavy Ion Collider (RHIC)	Heavy ions, protons	2 × (30-100) GeV/u 2 × (30-250) GeV
Massachusetts Institute of Technology Bates Linear Accelerator Center	Electrons	0.1-1 GeV
Michigan State University National Superconducting Cyclotron Laboratory	Light to very heavy ions	10-200 MeV/u
Indiana University Cyclotron Facility	Protons, light ions	100-500 MeV
Argonne National Laboratory (IL) Argonne Tandem Linac Accelerator System	Light to very heavy ions	0.3-20 MeV/u
Oak Ridge National Laboratory (TN) Holifield Radioactive Ion Beam Facility	Light to heavy ions	0.1-12 MeV/u
Lawrence Berkeley National Laboratory (CA) 88" Cyclotron	Protons, light to very heavy ions	1-55 MeV 1-35 MeV/u

beams up to an energy of 1 GeV. The pulsed linac and the isochronous recirculator provide currents in excess of 80 μA at a duty-factor of up to 1 percent. The existing accelerator-recirculator system feeds the recently completed South Hall Ring, which will provide close to 100 percent-duty-factor beams. The 190 m-circumference ring will operate in the energy range up to 1 GeV at peak circulating currents of up to 80 mA, and extracted currents will be up to 50 μA.

Three superconducting cyclotrons have been built at the National Superconducting Cyclotron Laboratory of Michigan State University (MSU/NSCL). Two

Technology	Research Areas
Superconducting accelerator Polarized-electron beams Three simultaneous target stations	Structure of hadrons Quark-gluon degrees of freedom in nuclei Electromagnetic response of nuclei
Colliding beams Polarized-proton beams Superconducting magnets	Quark-gluon plasma Hot compressed nucleonic matter Spin physics
Polarized-electron beams Electron stretcher/storage ring Internal targets	Fundamental symmetries and interactions Structure of hadrons and nuclei Spin structure of nucleons and nuclei
Superconducting cyclotrons Superconducting magnets Radioactive beams	Nuclear structure with radioactive beams Liquid-gas phase transition Nuclear astrophysics
Polarized, stored cooled beams Internal targets	Nucleon-nucleon/meson interactions Spin structure of nuclei Fundamental symmetries and chirality
Superconducting accelerator Selected radioactive beams	Nuclear structure at the limits Nuclear astrophysics with radioactive beams Ion trapping and fundamental symmetries
Two-accelerator ISOL facility Radioactive beams	Nuclear structure with radioactive beams Nuclear astrophysics Decay studies far off stability
ECR ion sources Rare-isotope beams	Nuclear structure at the limits Heavy-element research Atom trapping and fundamental symmetries

of these, the K500 and the K1200, are currently being coupled in a program to upgrade the capabilities of the MSU system. The K500, the world's first superconducting cyclotron, operated from 1982 to 1988 in support of the nuclear physics program at MSU. The K1200 is the world's highest-energy (~10 GeV) continuous-wave (CW) cyclotron and has been used in support of the nuclear physics program since 1988. Both the K500 and the K1200 operate at 5 T. The upgraded facility will provide intense, high-energy beams of heavy ions for in-flight fragmentation to produce intense secondary radioactive beams.

TABLE A.2 University Accelerators

Facility	Beam Characteristics: Species	Energies
Florida State University Tandem Linac	Protons, light to medium heavy ions	2-10 MeV/u
State University of New York at Stony Brook Tandem-Linac	Protons, light to medium heavy ions	2-10 MeV/u
University of Notre Dame (IN) Accelerator Facility	Protons, light to medium heavy ions	2-21 MeV 0.1-8 MeV/u
Texas A&M University Cyclotron Institute K500 Superconducting Cyclotron	Protons, light to heavy ions	2-70 MeV/u
University of Washington Tandem Linac	Protons, light to medium heavy ions	2-16 MeV 2-10 MeV/u
Triangle Universities Nuclear Laboratory (NC) Tandem Accelerator	Protons, light ions, neutrons, photons	1-10 MeV/u
University of Wisconsin Tandem Accelerator	Protons, light ions	2-12 MeV 1-7 MeV/u
Yale University (CT), Wright Nuclear Structure Laboratory Tandem Accelerator	Protons, light to heavy ions	1-40 MeV 1-15 MeV/u

The Indiana University Cyclotron Facility is active in areas of beam-cooling technologies and polarized proton beams. Two cyclotrons provide protons with energies up to 200 MeV for direct beams, as well as serving as the injector complex for the Cooler Ring. This ring can accelerate protons to an energy of 500 MeV. In addition, the Cooler Ring has a state-of-the-art electron-cooling system capable of providing high-resolution, very dense beams. A new synchrotron injector into the Cooler was recently completed, providing a two-orders-of-magnitude increase in beam intensity. Distinguishing characteristics of the facility include polarized beams in all machines and internal polarized gas-jet targets in the cooler.

Technology	Research Areas
Superconducting cavities Polarized lithium beam	Nuclear structure and decay Spin effects in nucleus-nucleus collisions
Superconducting cavities	Nuclear structure Heavy-ion reactions Atom trapping and spectroscopy
Radioactive beams Intense low-energy stable beams	Nuclear structure and reactions Fundamental symmetries Nuclear astrophysics
Intermediate-energy heavy ions Selected radioactive beams	Nuclear structure and reaction dynamics Nuclear astrophysics with radioactive beams Fundamental symmetries
Superconducting cavities Terminal ion source	Nuclear reactions with heavy ions Tests of fundamental symmetries
High-resolution light-ion beams Polarized beams	Fundamental symmetries Inter-nucleon reactions and light nuclei Nuclear astrophysics
Polarized beams	Few-body systems Fundamental symmetries Spin degrees of freedom in nuclei
High-resolution beams	Nuclear structure Heavy-ion reactions Nuclear astrophysics

ATLAS at Argonne National Laboratory consists of a superconducting linear accelerator, which is injected by either a 9-MV tandem Van de Graaff or a new positive-ion injector (PII) consisting of two ECR ion sources and a superconducting injector linac of novel design. Using the PII, ATLAS routinely accelerates intense beams up to uranium with energies above the Coulomb barriers and with excellent beam properties. The accelerator has a 100 percent duty cycle, can provide very short beam pulses (<150 psec), and is ideal for high-resolution heavy-ion nuclear physics research where nuclear structure effects are particularly important.

The Holifield Radioactive Ion Beam Facility (HRIBF) was recently brought

FIGURE A.1 Illustration and geographical distribution of nuclear physics laboratories in the United States. Shown are major national user facilities, as well as the smaller, dedicat-

SUNY at Stonybrook
Tandem-Linac

MIT/Bates
Accelerator

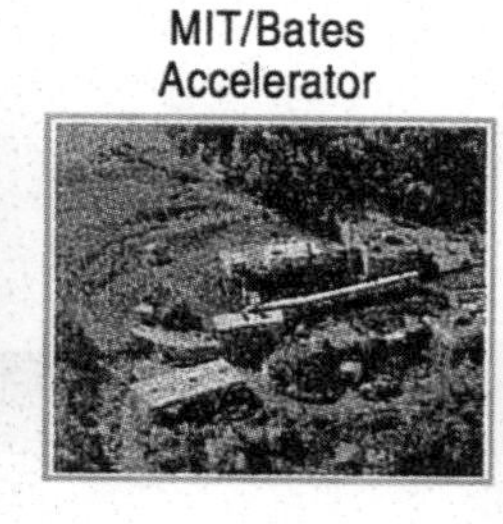

Yale University
ESTU Tandem Accelerator

Brookhaven National
Laboratory, RHIC

Thomas Jefferson National
Accelerator Facility,
CEBAF

Oak Ridge
National Laboratory
Holifield Radioactive
Ion Beam Facility

Florida State University
Tandem - Linac

Triangle Universities
National Laboratory
Durham

ed university laboratories that, together and in a synergistic relationship, cover the broad range of science that is described.

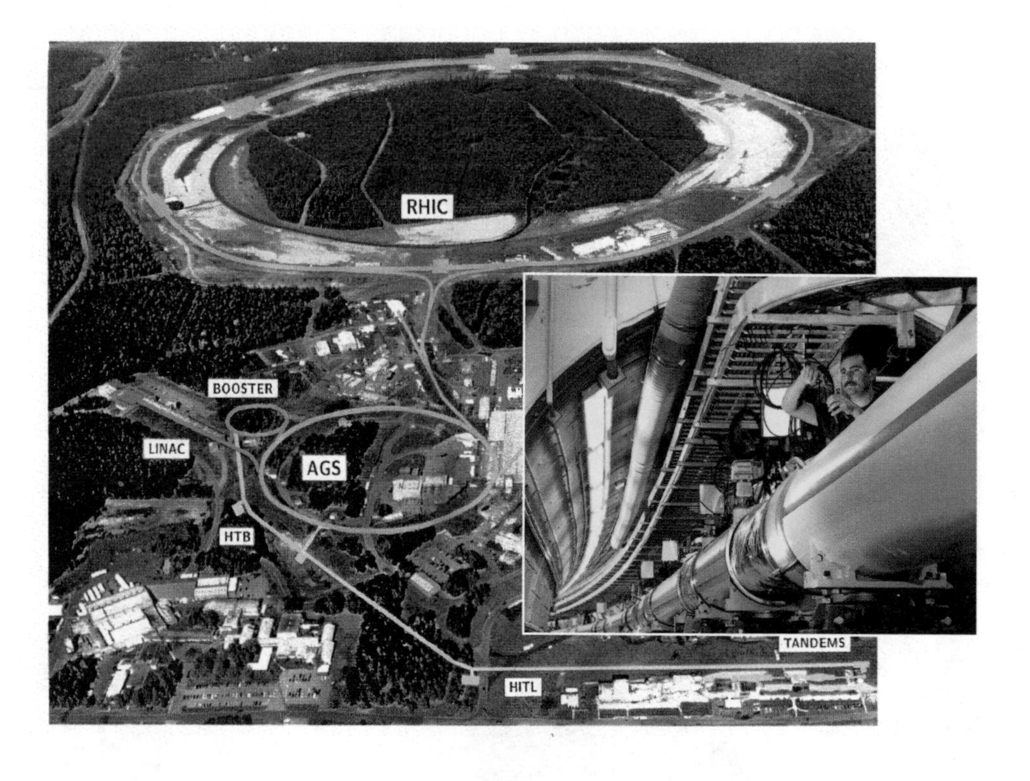

FIGURE A.2 Aerial view of the Relativistic Heavy Ion Collider (RHIC) accelerator complex at Brookhaven National Laboratory on Long Island. The facility will generate two counter-circulating, high-energy beams of heavy (for example, gold) nuclei; these beams will intersect and collide in six interaction regions located uniformly around the rings. The insert shows a recent photo of the tunnel that houses the accelerator, with an array of powerful superconducting magnets needed to bend the high-energy beams. (Courtesy RHIC.)

into operation at Oak Ridge National Laboratory to permit research with precisely controlled beams of radioactive nuclei generated by the ISOL (isotope separator online) technique. Light ions (protons, deuterons, and others) accelerated by the Oak Ridge Isochronous Cyclotron strike an isotope-production target. Radioactive atoms produced in the target are extracted, ionized, mass selected, charge exchanged, preaccelerated, and injected into a 25-MV electrostatic tandem for acceleration to final energies.

The 88" Cyclotron at Lawrence Berkeley National Laboratory is a variable-energy isochronous cyclotron with spiral-sector focusing. A new ECR source, operating at 14 GHz, and producing high intensities of highly charged ions has revolutionized the acceleration of heavy nuclei. The 88" Cyclotron now provides ion beams of energies up to and above the Coulomb barrier for medium-mass nuclei, as well as lighter ion beams with energies up to 35 MeV/nucleon.

On a considerably smaller scale, university accelerators serve the important function of providing specialized capabilities that cannot be met by the national user facilities. Examples of problems requiring lower beam energies and thus smaller accelerators are studies of fundamental symmetries in light nuclei, nuclear structure far from stability, reactions near the Coulomb barrier, or the study of nuclear reactions that are relevant to astrophysics.

At present, eight small accelerators (also shown in Figure A.1) are supported by NSF and DOE. These are situated on university campuses and serve local faculty and students, but they also are often used by visitors who take advantage of their unique capabilities. In most cases, the capabilities of these machines result from intensive, sometimes even speculative, development work on new instrumentation, such as novel detectors or ion sources. The modest cost of accelerator operation and the availability of the accelerator over long periods of time permit developments that could not be carried out in a cost-effective way on large machines.

The accelerators at Florida State University, the State University of New York at Stony Brook, and the University of Washington consist of tandem electrostatic accelerators coupled to superconducting linear accelerators to augment their energy. At Florida State University, polarized Li ions are produced by laser pumping and are used to study spin effects in elastic and inelastic processes. New capabilities at Stony Brook include the development of a magneto-optical trap for production and trapping of short-lived radioactive nuclei. The accelerator at the University of Washington is equipped with a new ion source in the tandem terminal to provide the intense beams of alpha particles required for weak interaction studies.

The early work on production of radioactive beams at the University of Notre Dame led to systematic studies with ^{8}Li of importance to cosmology. At Texas A&M University, a new beam-analysis system and spectrometer coupled to the K500 Superconducting Cyclotron provides radioactive-beam capabilities for studies of exotic nuclei and astrophysics. The availability of polarized beams of protons and deuterons at the Triangle Universities Nuclear Laboratory (TUNL) in North Carolina and at the University of Wisconsin is of importance to studies of few-body systems. The technical developments for polarized beams and targets are applied at the larger facilities in a symbiotic relationship between small university accelerators and the large national and international facilities. The large ESTU tandem accelerator at Yale University, which has been used to accelerate heavy ions of mass up to gold, is being augmented with a mass-separator facility in the terminal of the machine for use primarily in the nuclear-structure and nuclear-astrophysics programs.